I0817745

BISON
BOOKS

*Theodore Roosevelt's*

# Wilderness Writings

THEODORE ROOSEVELT

New Edition

*Edited and with an introduction by Paul Schullery*

UNIVERSITY OF NEBRASKA PRESS | LINCOLN

The University of Nebraska Press is part of a land-grant institution with campuses and programs on the past, present, and future homelands of the Pawnee, Ponca, Otoe-Missouria, Omaha, Dakota, Lakota, Kaw, Cheyenne, and Arapaho Peoples, as well as those of the relocated Ho-Chunk, Sac and Fox, and Iowa Peoples.

Library of Congress Cataloging-in-Publication Data

Names: Roosevelt, Theodore, 1858–1919, author. | Schullery, Paul, editor.
Title: Theodore Roosevelt's wilderness writings / Theodore
Roosevelt; edited and with an introduction by Paul Schullery.
Description: Lincoln: University of Nebraska Press,
2025. | Includes bibliographical references.
Identifiers: LCCN 2024018775
ISBN 9781496240521 (paperback)
ISBN 9781496242549 (epub)
ISBN 9781496242556 (pdf)
Subjects: LCSH: Nature conservation—United States—History. | Wilderness areas—United States—History. | Roosevelt, Theodore, 1858–1919—Political and social views. | Wildlife conservation—United States—History. | Environmental protection—United States—History. | Outdoor life—United States. | Naturalists—United States—Biography. | BISAC: HISTORY / United States / State & Local / General | TRAVEL / Special Interest / Adventure
Classification: LCC QH76 .R658 2025 | DDC 333.78/20973—dc23/eng/20240601
LC record available at https://lccn.loc.gov/2024018775

Designed and set in Questa by K. Andresen.

# Contents

**MAP**

# Editor's Note on the Text

THE TEXTS IN THESE CHAPTERS ARE TRANSCRIBED AS PRECISELY as possible from the original publications, with full respect to all of TR's choices in spelling, grammar, and style. Besides his own grammar/spelling/style choices and tastes, his numerous publishers and editors during the thirty years of these essays' original appearances would have differed one from another in their style requirements. This may give rise to minor style variations from chapter to chapter, but it is judged best to stick with the original texts, just as TR saw them printed.

All of the selections appear with their original titles except for chapter 1, which was the foreword to *A Book-Lover's Holiday in the Open*; chapter 10, which was an untitled speech; and chapter 13, which had the full title of "On the Cattle Ranges: The Pronghorn Antelope."

TR's use of common plant and animal names can be confusing or even outdated. A few examples make the point. He enjoyed romanticizing names, so a golden eagle became the "great war eagle," and on one occasion the bison became the "great wild ox." The pronghorn, though not strictly an antelope, was and still is often referred to as such, or as a pronghorn antelope, or a prongbuck, or just as an antelope. Likewise, the name "wapiti," which TR favored, still occasionally lingers in informal conversations about elk. Neither formal nor popular nomenclature are the result of a tidy process.

Last, some of his natural-history generalizations on landscapes and wildlife sweep a bit too broadly. Nowhere is TR revealed more as a man of his time than in the state of his expertise. As remarkable and progressive as his natural history views were for their time, it was inevitable that he couldn't be right all the time. Indeed, for just one example, with the advantage of more than 120 years' additional scientific study of Yellowstone Park since TR's time, I could easily write an essay of commentary, elaboration, and occasional correction as long as chapter 8, analyzing his account of his visit there. In his day he was a leader of the centuries-long process by which we have learned, unlearned, and relearned countless elements of our wilderness and its inhabitants. We read him today as we would read Audubon or any other historic observer of nature, not as an up-to-the-minute authority but as a gifted and historically influential naturalist caught up in a uniquely dynamic era in our ever-changing relationship with the natural world.

# Introduction

THEODORE ROOSEVELT (1858–1919), THE TWENTY-SIXTH PRESident of the United States (1901–9), remains one of the most colorful, complex, and controversial political figures in American history.[1] And no wonder. He was a true believer in free enterprise, yet a vigorous crusader against the trusts it spawned. He was an aggressive, even imperialistic, chief of state, yet was the first president to win the Nobel Prize for peace. He was a tireless champion of "manly virtues," but for him those virtues included the most refined pleasures of the fine arts and the wide realm of natural history. Perhaps the intellectual and spiritual intricacies of the man can be symbolically illuminated simply by pointing out that he was a church-going Darwinist.

In less than twenty years, from 1882 to 1901, TR rose through positions as New York State assemblyman, U.S. Civil Service commissioner, New York City police commissioner, assistant secretary of the navy, governor of New York, vice president, and president. But to appreciate the breadth of his activities and accomplishments, it's best to set aside that breathtaking political ascent for a moment and consider what else he fit into his busy days. He was an important American historian; his *The Naval War of 1812*, published when he was just twenty-three and still a valuable work on the subject, was followed by other notable histories and biographies, and this career was capped by a term as president of the American Historical Association (1912–13). His political essays, commentary, and polemics would have distinguished him in that field even had he held no public office. For added color we have his years as a North Dakota ranchman, when he served both as a deputy sheriff and as chairman of the near-vigilante Little Missouri River Stockmen's Association. And of course there was Colonel Roosevelt's crowded hour leading his Rough Riders through heavy fire to fame in Cuba.

All these compartments and chapters of TR's life are memorable, but it is likely that none of them approach the historical significance of yet another of his passions: everything to do with the natural world. Any summary of his unique achievements as a sportsman-naturalist-conservationist almost has to begin with a recitation of the lands he set aside in one form or another during his presidency, more than 230 million acres including 150 national forests (creating the now-venerable U.S. Forest Service in the process), 5 national parks, 18 national monuments, and 55 other bird and game preserves.[2]

More to the point of this book, there is also his enduring literary monument to consider. Many accounts of his hunting and wilderness experiences tend to list only the first three of his hunting books—*Hunting Trips of a Ranchman* (1883), *Ranch Life and the Hunting Trail* (1888), and *The Wilderness Hunter* (1893), among the foremost classics in that literary tradition. But he continued to publish wisely and eloquently about his experiences and nature studies, to a total of ten books of his own and five more of which he was coauthor or coeditor. All of these were rich in TR's invigorated blend of hunting adventure and lore, extensive natural-history observations, lectures on good sportsmanship, and pleas for the protection of wild places. In assembling this book I've drawn fairly on these less-known books, as well as on a few periodical publications.

Our ideas of wilderness, indeed of nature itself, have evolved and grown more complicated and frequently vexing since TR and his associates used those terms in what must have seemed to them intuitive, straightforward ways.[3] The American wilderness in 1901, when Roosevelt became president, was quantitatively and qualitatively unlike the "wilderness areas" we enjoy today. Most Americans still held a nearly biblical notion of wilderness as dangerous, possibly evil, and most certainly there to be conquered. There was nothing like the legislative and bureaucratic armature—starting with legislated boundaries—that exists around the places we have now formally designated as Wilderness with that all-important capital "W." There were only all those vast, beckoning stretches of land no one was putting to much use—lands many saw as going to waste.

In the years up to the passing of the Wilderness Act in 1964, and with rapidly increasing intensity ever since, we as a nation have engaged in an endless and often heated conversation over just what is meant by terms such as wilderness, wildness, nature, and natural (even the term "waste" has undergone thorough analysis and dissection). Such debates are vitally important to the future of the modern conservation movement, but they were beyond the thinking or interests of most people in TR's day. The perceptual steamroller of wilderness conquest had for good or ill already altered much of the American landscape and depleted or destroyed its wildlife, and TR and his friends could see it bearing down on what wild country was left. In some of the selections in this book he discusses the fate of the wildlands of his day, and especially of the wild inhabitants of those places.

Roosevelt enjoyed wilderness the way he did everything, with a vigor and intensity that will always make many people nervous. His writing displays that same energy. He was a romantic. He saw the glamor, the high adventure, and the glory in his wilderness experiences, and he did not deny himself the gratification of enthusiastic literary celebration. Historians who do not take these celebrations seriously are missing an important and revealing source of information about Roosevelt's character. By today's standards his vivid prose may seem overdone, but many of us still relish it as no less positive a reading experience than the fiery prose sermons of John Muir or the serene and timid meditations of John Burroughs. As a writer TR had Muir's gift for well-aimed polemic and Burroughs's eye for fine detail. And in one thing he far surpassed all the other writers: the stupendous size of his audience.

## ON READING ROOSEVELT TODAY

Whatever our areas of interest, reading our celebrated forebears is often a troubling exercise in getting uncomfortably close to these historic heroes. Who can read the writings of Thomas Jefferson, even just the Declaration of Independence, without marveling that this brilliant person, who so enduringly articulated the essence of human freedom, could at the same time own, breed, buy, and sell his fellow human beings? He may not have done this without thought or reservation, but that he could do it at all was still monstrous.

Likewise, TR rarely lets you forget he was a man of his time—in some ways among the most enlightened of Americans, yet in other ways proud citizen of a society whose bigotry and injustices we still struggle to overcome. His was above all a *man's* world. His exclusive use of "man" to represent all humanity, and of "he/him/his" when generalizing about people, grate on the modern ear. We are infuriated by his references to "savages" and other racist remarks, as well as by his blind pride in his own "masterful" race and his apparent conviction there is really only one way to be "civilized."

No excuses: Theodore Roosevelt, though blessed with "the richest and most varied background any individual had brought to the Presidency since the time of Thomas Jefferson,"[4] was complicated and compromised well within the prevailing range of his contemporaries. His principles and prejudices, wisdoms and idiocies all were rattling around in that square skull and driving him to be as good a man of his times as he knew how to be. That's not good enough for us today, any more than we will necessarily measure up to the ideals of our descendants in another hundred years, but it was good enough in his time for us to wish to pay close attention to how he made his way there.

What's more, to further bewilder our perspectives, TR is a looming presence in another historic controversy: he was a passionate sport hunter. For many years following his death, historians largely neglected TR's extraordinary career in both nature study and conservation in good part because too many people either disapproved of or just didn't know what to make of his enthusiasm for hunting. Historian Edward Wagenknecht, whose *The Seven Worlds of Theodore Roosevelt* (1958) is still a good introduction to the breadth of TR's life pursuits, had to admit that he couldn't understand how "a man who loved animals as much as Roosevelt did could still enjoy *killing* them."[5] I'm pleased to say that scholarship about hunting in general and about TR's hunting in particular has come a long way since such simplistic questions about a cultural act as intellectually, emotionally, and morally convoluted as sport hunting could be asked with so little apparent interest on the part of the questioners in trying to find the answers. Scholars passed their judgment and dismissed TR as a bloodthirsty cowboy. Getting to know the hunting culture as portrayed in this book might not change your mind or your moral stance on TR's behavior and convictions, but it will at least inform you that TR was not acting without a moral stance of his own.

In fact, I'm especially pleased that this book does provide—all to TR's credit, not to mine—an impressive display, a model in its day in fact, of how an expert naturalist integrated into his wilderness days the various pursuits that go into earning a full acquaintance with wild places. TR was a wilderness multitasker. A vivid hunting tale might be interrupted to consider and celebrate the song, behavior, and general beauty of a favorite bird encountered along the trail, and then conscientiously concluded with a report on the feeding habits of the animal just shot, gained from an immediate field necropsy and perhaps enriched by invoking the books of other wildlife authorities. Also, somewhere in, before, or following this little lesson in the hunter-naturalist's wide-ranging interests, there might also appear an exhortation to *all* sportsmen and non-sportsmen to learn as much as they can about any animal they observe or pursue, or on the urgency of preserving the habitats of all the wild plants and animals.

For TR, all these were matters of personal responsibility. In his view the true sportsman sought much more than dead meat. The very act of hunting was wrapped up in a swirl of interrelated obligations and rewards that were themselves wrapped up in his definition of good citizenship. The readings that follow will give you a fair demonstration of how this one prominent citizen approached nature on a variety of fronts, all connected, all profoundly important to him.

## THE SELECTIONS

Part 1, "Wilderness Adventures," begins with "The Joy of the Wild," a stage-setter for all the rest. Here is adventurer TR at his most purely celebratory, deploying his most heroic, proto-machismo prose—

his particular brand of *sturm und drang* sermonizing. At its heart is exhortation—go to the wild country!—backed by a flood of vivid imagery of all that awaits you "among untrodden forests, in the swirl of wild waters, and in the blast of snow blizzard or thunder-shattered hurricane." But then in the midst of all this death-defying adventure, TR's own complicated personal code asserts itself and he reminds us: *Don't forget to take your books!*

Chapter 2, "The American Wilderness: Wilderness Hunters and Wilderness Game," is an extended elaboration on the same themes but in more of TR's voices. There's the same vigorous prose, but now it's the historian, the ranchman, and the hunter-naturalist taking their turns weaving a saga of continental exploration as geographical and zoological discovery, always underlaid with his conviction of the inevitability of American nation-building. TR really knew how to put the "man" in "manifest destiny."

Chapter 3, "The Big-Horn Sheep," puts all this stage-setting to work on one magnificent element of wilderness life. Like several of TR's accounts of North American big game, this essay was regarded as a model of the form. George Bird Grinnell, pioneering American conservationist (widely referred to as the father of Glacier National Park) and hunter-naturalist, described this essay as the "best published account" of the animals' life history.[6] It's also a perfect example of TR's intertwining of sport and nature study. One studied natural history to become a better hunter, just as one hunted to become a better naturalist.

Because of media and public attention, Roosevelt found hunting difficult during his presidency. His attempt to hunt bears in Mississippi in 1902 was an embarrassing fiasco: "My kind hosts, with best of intentions, insisted upon turning the affair into a cross between a hunt and a picnic, which always results in a failure for the hunt and usually a failure for the picnic."[7] By 1907 he had learned to manage these problems, and the result was the hunt described in chapter 4, "In the Louisiana Cane-Brakes," which features his trademark broad-ranging cultural and natural-history asides, among the most important being descriptions of two legendary old bear hunters, Ben Lilly and Holt Collier, the latter a former slave and Confederate veteran; and his enviable sighting of the spectacular now-nearly-extinct ivory-billed woodpecker. The bear hunt took place in the Old

South's bear hunting tradition, with a group of mounted men on horseback following a pack of hounds in a wild and woolly version of the traditional English fox hunt.

In the winter of 1913–14, following his presidency and his epic African safari, TR participated in what was certainly the riskiest of his travels, described in chapter 5. "Down an Unknown River," is the tale of a jungle exploration down an unmapped tributary of the Amazon system. It is impossible to overstate the dangers and hardship his party endured; the trip was simply too much for the aging TR, and both he and his son Kermit nearly died from exposure, disease, starvation, or injuries. As if the many miles of wild, treacherous rapids and countless other jungle misfortunes weren't enough, one of the party's boatmen murdered another, which resulted in one of the most astonishingly unlikely scenes in the history of the American presidency: a former president of the United States in his mid-fifties, rifle at the ready, moving warily with an unarmed companion from the dead man's corpse through heavy jungle foliage along the murderer's trail.[8]

Part 2, "Wilderness Preservation," begins with TR's forceful appeal, published in 1912, for the creation of an agency to manage the nation's growing number of national parks. Following the establishment of Yellowstone in 1872, there was a gradual accumulation of national parks, most with their own peculiar set of management arrangements. Lacking direction and advocacy, the parks were administrative orphans, and once it was clear that these new institutions were not only going to last but had immense potential for public good, TR was loud in the chorus of friends of the parks making the case that the parks' business be conducted on better terms. The National Park Service was finally established in 1916.

John Muir was among America's greatest champions of wilderness protection. Though he has been most closely identified with the California Sierra, the reach of his theories and teachings was in fact far broader than any one region or country. In chapter 7 TR describes their only direct encounter, when Muir and the then-president camped in Yosemite National Park. The two men were hardly in complete agreement about wilderness lands, TR being much more tolerant than Muir of the nation's need for aggressive use of vast reaches of public domain. But they shared a belief in the need to protect wilderness lands such as those in the Yosemite. Generations of us would give much to have been a silent audience to what must have been some wonderful conversations along the trail and camping under the big trees.

At the beginning of chapter 8, "Wilderness Reserves: The Yellowstone Park," TR lays out his own view of the need for comprehensive, landscape-scale protection of wild lands. His concern for preserving wild game species, his conviction that such protection is at heart a democratic necessity, and his interest in having some such lands kept in "a state of nature" are all at play here, leading up to his account of his extended presidential visit to Yellowstone in 1903. On this occasion, because sport hunting had been banned in the park some twenty years earlier, TR the naturalist took charge, relishing the almost constant opportunities to observe and study the park's remarkable array of wild mammals. It was a trip enviable for its grand isolation; this early in the season the park was not yet open to the public, and good portions of TR's time were spent in deep snow in the company only of John Burroughs and a few park officials.[9]

TR's passion for birds of all kinds and sizes is showcased in chapter 9, "Bird Reserves at the Mouth of the Mississippi." Pelican Island was the first (1903) of fifty-one bird reserves he created during his presidency, a less-known but vitally essential part of his conservation legacy. Here TR divides his praise between the local protectors of these fabulous sanctuaries and the state and national agencies and organizations that supported them, concentrates his scorn on the "sordid bird-butchers" who committed almost unimaginable slaughter on the beautiful birds of the Gulf Coast, and saves his warmest admiration for the spectacle of the birds themselves.

TR's brief visit to the Grand Canyon of the Colorado River, represented by his off-the-cuff remarks in chapter 10, is best remembered for his exhortation/appeal to "Leave it as it is," which became an enduring catchphrase and slogan for defenders of wild beauty. In 1908 he designated the Grand Canyon a national monument, and in 1919 President Wilson elevated it to national park status.

Part 3, "Natural History," opens with "Small Country Neighbors," TR's happy account of the wild creatures he observed and enjoyed near his various homes during his presidency. Among the many tales of this or that species of mammal or bird and the diversions into discussions of various landscapes and habitats, we find another off-the-charts-enviable sighting. In May 1907 TR had a reasonably good sighting, one of the last in the wild, of a small flock of passenger pigeons, a bird that once flew in gigantic flocks across the eastern United States but which had been pushed to the very edge of extinction by the time of TR's sighting. Indeed, the last known passenger pigeon, named Martha, died a captive at a Cincinnati zoo in 1914.[10]

Chapter 12, "The Wapiti or Round-horned Elk," is TR's tribute to what was almost certainly his favorite species of big game. Though he preferred the traditional Native American name "wapiti," he seemed resigned to the reality that most people were going to call it an elk. He gives us what for the time was an excellent summary of elk life history, following it with some of hunting experiences which, as usual, drift over into more notes on the habits and behavior of the species.

TR's political opponents sometimes referred to him as a "cowboy," or, if especially annoyed at him, "that damned cowboy," but historians remind us that he was not a cowboy; he was a *ranchman*, a different creature by virtue of social distinction. During his various times managing his North Dakota ranch he lived with cowboys and performed all the duties they did, but it was his ranch and they were his cowboys. In chapter 13, a memoir of those times in the 1880s, TR gives us his personal take on times and places just then being memorialized in fiction by Owen Wister and in paintings and sculptures by Frederic Remington. Predictably, he finds room in the story for a pronghorn hunt, followed just as predictably by several pages on the birds of the prairie (with a brief side trip to Tennessee on behalf of the mockingbird).

Chapter 14, "My Life as a Naturalist," rounds out this collection. It is a memoir of his childhood fascination with nature so intense that it nearly set him on a life course as a professional naturalist. It wasn't until he was a Harvard undergraduate that he became so disillusioned with what he saw as the increased domination of laboratory science over traditional nature study that he abandoned the career of a full-time naturalist and turned to the remarkable life for which we now remember him.

## NOTES

1. Though there are countless biographical works related to TR, the prevailing one-author biography is Edmund Morris's magisterial three-volume work, *The Rise of Theodore Roosevelt* (New York: Coward, McCann & Geoghegan, 1979); *Theodore Rex* (New York: Random House, 2001); and *Colonel Roosevelt* (New York: Random House, 2010). From Morris's references and citations the interested reader can explore the immense scholarly literature devoted to TR. All students and enthusiasts of TR's life and various careers would do well to join The Theodore Roosevelt Association, PO Box 719, Oyster Bay, NY 11771, theodoreroosevelt.org, publisher for the past half century of the venerable peer-reviewed *Theodore Roosevelt Association Journal*. The TRA is a fine gateway into a world of TR-related institutions, research collections, and other helpful services.

2. Three milestone works from the cornerstone of our understanding of TR's career as a naturalist and conservationist: Paul Cutright's *Theodore Roosevelt, The Naturalist* (New York: Harper & Brothers, 1956), and *Theodore Roosevelt, The Making of a Conservationist* (Urbana: University of Illinois Press, 1985), and Douglas Brinkley's admirably exhaustive *The Wilderness Warrior: Theodore Roosevelt and the Crusade for America* (New York: Harper Perennial, 2009).

3. The scholarly literature of the history and meaning of the wilderness is substantial, and the corresponding popular literature is immense, but for three generations many readers seeking a grounding in the evolution of the idea of wilderness have turned first to Roderick F. Nash's nearly venerable *Wilderness and the American Mind* (New Haven: Yale University Press, 5th ed., 2014). For a much more extended study of how humans have related to nature, you might want to start with Clarence J. Glacken, *Traces on the Rhodian Shore: Nature and Culture in Western Thought from Ancient Times to the End of the Eighteenth Century* (Berkeley: University of California Press, 1967).

4. E. C. Blackorby, "Theodore Roosevelt's Conservation Policies and Their Impact Upon America and the American West," *North Dakota History*, October 1958, 108.

5. Edward Wagenknecht, *The Seven Worlds of Theodore Roosevelt* (New York: Longman, Green, 1958), 20. My own entry into formal discussions of the controversy surrounding TR's hunting is "Theodore Roosevelt: The Scandal of the Hunter as Nature Lover," in *Theodore Roosevelt, Many-Sided American*, ed. Natalie A. Naylor, Douglas Brinkley, and John Allen Gable (Interlaken, NY: Hofstra University and Heart of the Lakes Publishing, 1992), 221–30.

6. George Bird Grinnell, "President Roosevelt as a Sportsman," *Forest and Stream*, December 5, 1903, 437.

7. Theodore Roosevelt, letter to Philip Stewart, in *The Letters of Theodore Roosevelt*, vol. 3, *The Square Deal, 1901–1903*, ed. Elting Morison (Cambridge: Harvard University Press, 1951), 387–89.

8. A detailed account of TR's Brazil adventure appears in Candice Millard's *The River of Doubt: Theodore Roosevelt's Darkest Journey* (New York: Broadway Books, 2005).

9. Besides the accounts in Cutright and Brinkley, cited above, concentrated studies of TR's visits to and relationship with Yellowstone National Park include my "A Partnership in Conservation . . . Roosevelt & Yellowstone," *Montana: The Magazine of Western History* 28, no. 3 (1978): 2–15, and Lee Whittlesey and Paul Schullery, "The Roosevelt Arch: A Centennial History of An American Icon," *Yellowstone Science* 11, no. 3 (2003): 2–24.

10. For a thorough analysis of the strong reliability of TR's sighting, see Alton A. Lindsey, "Was Theodore Roosevelt the Last to See Wild Passenger Pigeons?" *Proceedings of the Indiana Academy of Science* 86 (1976): 349–56.

THEODORE ROOSEVELT'S

# Wilderness Writings

PART 1

# WILDERNESS ADVENTURES

# 1

# The Joy of the Wild

THE MAN SHOULD HAVE YOUTH AND STRENGTH WHO SEEKS adventure in the wide, waste spaces of the earth, in the marshes, and among the vast mountain masses, in the northern forests, amid the steaming jungles of the tropics, or on the deserts of sand or of snow. He must long greatly for the lonely winds that blow across the wilderness, and for sunrise and sunset over the rim of the empty world. His heart must thrill for the saddle and not for the hearthstone. He must be helmsman and chief, the cragsman, the rifleman, the boat steerer. He must be the wielder of axe and of paddle, the rider of fiery horses, the master of the craft that leaps through white water. His eye must be true and quick, his hand steady and strong. His heart must never fail nor his head grow bewildered, whether he face brute and human foes, or the frowning strength of hostile nature, or the awful fear that grips those who are lost in trackless lands. Wearing toil and hardship shall be his; thirst and famine he shall face, and burning fever. Death shall come to greet him with poison-fang or poison-arrow, in shape of charging beast or of scaly things that lurk in lake and river; it shall lie in the swirl of wild waters, and in the blast of snow blizzard or thunder-shattered hurricane.

Not many men can with wisdom make such a life their permanent and serious occupation. Those whose tasks lie along other lines can lead it for but a few years. For them it must normally come in the hardy vigor of their youth, before the beat of the blood has grown sluggish in their veins.

Nevertheless, older men also can find joy in such a life, although in their case it must be led only on the outskirts of adventure, and although the part they play therein must be that of the onlooker rather than that of the doer. The feats of prowess are for others. It is for other men to face the peril of unknown lands, to master unbroken horses, and to hold their own among their fellows with bodies of supple strength. But much, very much, remains for the man who has "warmed both hands before the fire of life," and who, although he loves the great cities, loves even more the fenceless grassland, and the forest-clad hills.

The grandest scenery of the world is his to look at if he chooses; and he can witness the strange ways of tribes who have survived into an alien age from an immemorial past, tribes whose priests dance in honor of the serpent and worship the spirits of the wolf and the bear. Far and wide, all the continents are open to him as they never were to any of his fore-fathers; the Nile and the Paraguay are easy of access, and the borderland between savagery and civilization; and the veil of the past has been lifted so that he can dimly see how, in time immeasurably remote, his ancestors—no less remote—led furtive lives among uncouth and terrible beasts, whose kind has perished utterly from the face of the earth. He will take books with him as he journeys; for the keenest enjoyment of the wilderness is reserved for him who enjoys also the garnered wisdom of the present and the past. He will take pleasure in the companionship of the men of the open; in South America, the daring and reckless horsemen who guard the herds of the grazing country, and the dark-skinned paddlers who guide their clumsy dugouts down the dangerous equatorial rivers; the white and red and half-breed hunters of the Rockies, and of the Canadian woodland; and in Africa the faithful black gunbearers who have stood steadily at his elbow when the lion came on with coughing grunts, or when the huge mass of the charging elephant burst asunder the vine-tangled branches.

The beauty and charm of the wilderness are his for the asking, for the edges of the wilderness lie close beside the beaten roads of present travel. He can see the red splendor of desert sunsets, and the unearthly glory of the afterglow on the battlements of desolate mountains. In sapphire gulfs of ocean he can visit islets, above which the wings of myriads of sea-fowl make a kind of shifting cuneiform script in the air. He can ride along the brink of the stupendous cliff-walled canyon, where eagles soar below him, and cougars make their lairs on the ledges and harry the big-horned sheep. He can journey through the northern forests, the home of the giant moose, the forests of fragrant and murmuring life in summer, the iron-bound and melancholy forests of winter.

The joy of living is his who has the heart to demand it.

# 2

# The American Wilderness

## *Wilderness Hunters and Wilderness Game*

MANIFOLD ARE THE SHAPES TAKEN BY THE AMERICAN WILDERNESS. In the east, from the Atlantic Coast of the Mississippi Valley, lies a land of magnificent hardwood forest. In endless variety and beauty, the trees cover the ground, save only where they have been cleared away by man, or where toward the west the expanse of the forest is broken by fertile prairies. Toward the north, this region of hardwood trees merges insensibly into the southern extension of the great sub-arctic forest; here the silver stems of birches gleam against the sombre background of coniferous evergreens. In the southeast again, by the hot, oozy coasts of the South Atlantic and the Gulf, the forest becomes semi-tropical; palms wave their feathery fronds, and the tepid swamps teem with reptile life.

Some distance beyond the Mississippi, stretching from Texas to North Dakota, and westward to the Rocky Mountains, lies the plains country. This is a region of light rainfall, where the ground is clad with short grass, while cottonwood trees fringe the courses of the winding plains streams; streams that are alternately turbid torrents and mere dwindling threads of water. The great stretches of natural pasture are broken by gray sage-brush plains, and tracts of strangely shaped and colored Bad Lands; sun-scorched wastes in summer, and in winter arctic in their iron desolation. Beyond the plains rise the Rocky Mountains, their flanks covered with coniferous woods; but the trees are small, and do not ordinarily grow very closely together. Toward the north the forest becomes denser, and the peaks higher; and glaciers creep down toward the valleys from the fields of everlasting snow. The brooks are brawling, trout-filled torrents; the swift rivers foam over rapid and cataract, on their way to one or the other of the two great oceans.

Southwest of the Rockies evil and terrible deserts stretch for leagues and leagues, mere waterless wastes of sandy plain and barren mountain, broken here and there by narrow strips of fertile ground. Rain rarely falls, and there are no clouds to dim the brazen sun. The rivers run in deep canyons, or are swallowed by the burning sand; the smaller water-courses are dry throughout the greater part of the year.

Beyond this desert region rise the sunny Sierras of California, with their flower-clad slopes and groves of giant trees; and north of them along the coast, the rain-shrouded mountain chains of Oregon and Washington, matted with the towering growth of the mighty evergreen forest.

The white hunters, who from time to time first penetrated the different parts of this wilderness, found themselves in such hunting grounds as those wherein, long ages before, their Old-World forefathers had dwelled; and the game they chased was much the same as that their lusty barbarian ancestors followed, with weapons of bronze and of iron, in the dim years before history dawned. As late as the end of the seventeenth century the turbulent village nobles of Lithuania and Livonia hunted the bear, the bison, the elk, the wolf, and the stag, and hung the spoils in their smoky wooden palaces; and so, two hundred years later, the free hunters of Montana, in the interludes between hazardous mining quests and bloody Indian campaigns, hunted game almost or quite the same in kind, through cold mountain forests surrounding the Yellowstone and Flathead lakes, and decked their log cabins and ranch houses with the hides and horns of the slaughtered beasts.

Zoologically speaking, the north temperate zones of the Old and New Worlds are very similar, differing from one another much less than they do from the various regions south of them, or than these regions differ among themselves. The untrodden American wilderness resembles both in game and physical character the forests, the mountains, and the steppes of the Old World as it was at the beginning of our era. Great woods of pine and fir, birch and beech, oak and chestnut; streams where the chief game fish are spotted trout and silvery salmon; grouse of various kinds as the most common game birds; all these the hunter finds as characteristic of the New World as of the Old. So it is with most of the beasts of the chase, and so also with the fur-bearing animals that furnish to the trapper alike his life work and his means of livelihood. The bear, wolf, bison, moose, caribou, wapiti, deer, and big-horn, the lynx, fox, wolverine, sable, mink, ermine, beaver, badger, and otter of both worlds are either identical or more or less closely kin to one another. Sometimes of the two forms, that found in the Old World is the larger. Perhaps more often the reverse is true, the American beast being superior in size. This is markedly the case with the wapiti, which is merely a giant brother of the European stag, exactly as the fisher is merely a very large cousin of the European sable or marten. The extraordinary prong-buck, the only hollow-horned ruminant which sheds its horns annually, is a distant representative of the Old-World antelopes of the steppes; the queer white antelope-goat has for its nearest kinsfolk certain Himalayan species. Of the animals commonly known to our hunters and trappers, only a few, such as the cougar, peccary, raccoon, possum (and among birds the wild turkey), find their nearest representatives and type forms in tropical America.

Of course this general resemblance does not mean identity. The differences in plant life and animal life, no less than in the physical features of the land, are sufficiently marked to give the American wilderness a character distinctly its own. Some of the most characteristic of the woodland animals, some of those which have most vividly impressed themselves on the imagination of the hunters and pioneer settlers, are the very ones which have no Old-World representatives. The wild turkey is in every way the king of American game birds. Among the small beasts the coon and the possum are those which have left the deepest traces in the humbler lore of the frontier; exactly as the cougar—usually under the name of panther or mountain lion—is a favorite figure in the wilder hunting tales. Nowhere else is there anything to match the wealth of the eastern hardwood forests, in number, variety, and beauty of trees; nowhere else is it possible to find conifers approaching in size the giant redwoods and sequoias of the Pacific slope. Nature here is generally on a larger scale than in the Old-World home of our race. The lakes are like inland seas, the rivers, like arms of the sea. Among stupendous mountain chains there are valleys and canyons of fathomless depth and incredible beauty and majesty. There are tropical swamps, and sad, frozen marshes; deserts and Death Valleys, weird and evil, and the strange wonderland of the Wyoming geyser region. The waterfalls are rivers rushing over precipices; the prairies seem without limit, and the forest never ending.

At the time when we first became a nation, nine-tenths of the territory now included within the limits of the United States was wilderness. It was during the stirring and troubled years immediately preceding the outbreak of the Revolution that the most adventurous hunters, the vanguard of the hardy army of pioneer settlers, first crossed the Alleghanies, and roamed far and wide through the lonely, danger-haunted forests which filled the No-man's-land lying between the Tennessee and the Ohio. They waged ferocious warfare with Shawnee and Wyandot and wrought huge havoc among the herds of game with which the forest teemed. While the first Continental Congress was still sitting, Daniel Boone, the archetype of the American hunter, was leading his bands of tall backwoods riflemen to settle in the beautiful country of Kentucky, where the red and the white warriors strove with such obstinate rage that both races alike grew to know it as "the dark and bloody ground."

Boone and his fellow-hunters were the heralds of the oncoming civilization, the pioneers in that conquest of the wilderness which has at last been practically achieved in our own day. Where they pitched their camps and built their log huts or stockaded hamlets, towns grew up, and men who were tillers of the soil, not mere wilderness wanderers, thronged in to take and hold the land. Then, ill-at-ease among the settlements for which they had themselves made ready the way, and fretted even by the slight restraints of the rude and uncouth semi-civilization of the border, the restless hunters moved onward into the yet unbroken wilds where the game dwelled and the red tribes marched forever to war and hunting. Their untamable souls ever found something congenial and beyond measure attractive in the lawless freedom of the lives of the very savages against whom they warred so bitterly.

Step by step, often leap by leap, the frontier of settlement was pushed westward; and ever from before its advance fled the warrior tribes of the red men and the scarcely less intractable array of white Indian fighters and game hunters. When the Revolutionary War was at its height, George Rogers Clark, himself a mighty hunter of the old backwoods type, led his handful of hunter-soldiers to the conquest of the French towns of the Illinois. This was but one of the many notable feats of arms performed by the wild soldiery of the backwoods. Clad in their fringed and tasseled hunting-shirt of buckskin or homespun, with coonskin caps and deer-hide leggings and moccasins, with tomahawk and scalping-knife thrust into their bead-worked belts, and long rifles in hand, they fought battle after battle of the most bloody character, both against the Indians, as at the Great Kanawha, at the Fallen Timbers, and at Tippecanoe, and against more civilized foes, as at King's Mountain, New Orleans, and the River Thames.

Soon after the beginning of the present century Louisiana fell into our hands, and the most daring hunters and explorers pushed through the forests of the Mississippi Valley to the great plains, steered across these vast seas of grass to the Rocky Mountains, and then through their rugged defiles onward to the Pacific Ocean. In every work of exploration, and in all the earlier battles with the original lords of the western and southwestern lands, whether Indian or Mexican, the adventurous hunters played the leading part; while close behind came the swarm of hard, dogged, border-farmers,—a masterful race, good fighters and good breeders, as all masterful races must be.

Very characteristic in its way was the career of quaint, honest, fearless Davy Crockett, the Tennessee rifleman and Whig Congressman, perhaps the best shot in all our country, whose skill in the use of his favorite weapon passed into a proverb, and who ended his days by a hero's death in the ruins of the Alamo. An even more notable man was another mighty hunter, Houston, who when a boy ran away to the Indians; who while still a lad returned to his own people to serve under Andrew Jackson in the campaigns which that greatest of all the backwoods leaders waged against the Creeks, the Spaniards, and the British. He was wounded at the storming of one of the strongholds of the Red Eagle's doomed warriors, and returned to his Tennessee home to rise to high civil honor, and become the foremost man of his State. Then, while Governor of Tennessee, in a sudden fit of moody anger, and of mad longing for the unfettered life of the wilderness, he abandoned his office, his people, and his race, and fled to the Cherokees beyond the Mississippi. For years he lived as one of their chiefs; until one day, as he lay in ignoble ease and sloth, a rider from the south, from the rolling plains of the San Antonio and Brazos, brought word that the Texans were up, and in doubtful struggle striving to wrest their freedom from the lancers and carbineers of Santa Anna. Then his dark soul flamed again into burning life; riding by night and day he joined the risen Texans, was hailed by them as a heaven-sent leader, and at the San Jacinto led them on to the overthrow of the Mexican host. Thus the stark hunter, who had been alternately Indian fighter and Indian chief, became the President of the new Republic, and, after its admission into the United States, a Senator at Washington; and, to his high honor, he remained to the end of his days stanchly loyal to the flag of the Union.

By the time that Crockett fell, and Houston became the darling leader of the Texans, the typical hunter and Indian fighter had ceased to be a backwoodsman; he had become a plainsman, or a mountain-man; for the frontier, east of which he never willingly went, had been pushed beyond the Mississippi. Restless, reckless, and hardy, he spent years of his life in the lonely wanderings through the Rockies as a trapper; he guarded the slowly moving caravans, which for purposes of trade journeyed over the dangerous Santa Fé trail; he guided the large parties of frontier settlers who, driving before them their cattle, with all their household goods in their white-topped wagons, spent perilous months and seasons on their weary way to Oregon to California. Joining in bands, the stalwart, skin-clad riflemen waged ferocious war on the Indians, scarcely more savage than themselves, or made long raids for plunder and horses against the outlying Mexican settlements. The best, the bravest, the most modest of them all was the renowned Kit Carson. He was not only a mighty hunter, a daring fighter, a finder of trails, and maker of roads through the unknown, untrodden wilderness, but also a real leader of men. Again and again he crossed and recrossed the continent, from the Mississippi to the Pacific; he guided many of the earliest military and exploring expeditions of the United States Government; he himself led the troops in victorious campaigns against Apache and Navahoe; and in the Civil War he was made a colonel of the Federal Army.

After him came many other hunters. Most were pure-blooded Americans, but many were Creole Frenchmen, Mexicans, or even members of the so-called civilized Indian tribes, notably the Delawares. Wide were their wanderings, many their strange adventures in the chase, bitter their unending warfare with the red lords of the land. Hither and thither they roamed from the desolate, burning deserts of the Colorado to the grassy plains of the Upper Missouri; from the rolling Texas prairies, bright beneath their sunny skies, to the high snow peaks of the northern Rockies, or the giant pine forests and soft rainy weather, of the coasts of Puget Sound. Their main business was trapping, furs being the only articles yielded by the wilderness, as they knew it, which were both valuable and portable. These early hunters were all trappers likewise and, indeed, used their rifles only to procure meat or repel attacks. The chief of the fur-bearing animals they followed was the beaver, which abounded in the streams of the plains and mountains; in the far north they also trapped otter, mink, sable, and fisher. They married squaws from among the Indian tribes with which they happened for the moment to be at peace; they acted as scouts for the United States troops in their campaigns against the tribes with which they happened to be at war.

Soon after the Civil War the life of these hunters, taken as a class, entered on its final stage. The Pacific Coast was already fairly well settled, and there were few mining camps in the Rockies; but most of this Rocky Mountain region, and the entire stretch of plains country proper, the vast belt of level or rolling grass land lying between the Rio Grande and the Saskatchewan, still remained primeval wilderness, inhabited only by roving hunters and formidable tribes of Indian nomads, and by the huge herds of game on which they preyed. Beaver swarmed in the streams and yielded a rich harvest to the trapper; but trapping was no longer the main-stay of the adventurous plainsmen. Foremost among the beasts of the chase, on account of its numbers, its size, and its economic importance, was the bison or American buffalo; its innumerable multitudes darkened the limitless prairies. As the transcontinental railroads were pushed toward completion, and the tide of settlement rolled onward with ever-increasing rapidity, buffalo robes became of great value. The hunters forthwith turned their attention mainly to the chase of the great clumsy beasts, slaughtering them by hundreds of thousands for their hides; sometimes killing them on horse-back, but more often on foot, by still-hunting, with the heavy long range Sharp's rifle. Throughout the fifteen years during which this slaughter lasted, a succession of desperate wars was waged with the banded tribes of the Horse Indians. All the time, in unending succession, long trains of big white-topped wagons crept slowly westward across the prairies, marking the steady oncoming of the frontier settlers.

By the close of 1883 the last buffalo herd was destroyed. The beaver were trapped out of all the streams, or their numbers so thinned that it no longer paid to follow them. The last formidable Indian war had been brought to a successful close. The flood of the incoming whites had risen over the land; tongues of settlement reached from the Mississippi to the Rocky Mountains, and from the Rocky Mountains to the Pacific. The frontier had come to an end; it had vanished. With it vanished also the old race of wilderness hunters, the men who spent all their days in the lonely wilds, and who killed game as their sole means of livelihood. Great stretches of wilderness still remained in the Rocky Mountains, and here and there in the plains country, exactly as much smaller tracts of wild land are to be found in the Alleghanies and northern New York and New England; and on these tracts occasional hunters and trappers still linger; but as a distinctive class, with a peculiar and important position in American life, they no longer exist.

There were other men besides the professional hunters, who lived on the borders of the wilderness, and followed hunting, not only as a pastime, but also as yielding an important portion of their subsistence. The frontier farmers were all hunters. In the Eastern backwoods, and in certain places in the West, as in Oregon, these adventurous tillers of the soil were the pioneers among the actual settlers; in the Rockies their places were taken by the miners, and on the great plains by the ranchmen and cowboys, the men who lived in the saddle, guarding their branded herds of horses and horned stock. Almost all of the miners and cowboys were obliged on occasions to turn hunters.

Moreover, the regular army which played so important a part in all the later stages of the winning of the West produced its full share of mighty hunters. The later Indian wars were fought principally by the regulars. The West Point officer and his little company of trained soldiers appeared abreast of the first hardy cattlemen and miners. The ordinary settlers rarely made their appearance until in campaign after campaign, always inconceivably wearing and harassing, and often very bloody in character, the scarred and tattered troops had broken and overthrown the most formidable among the Indian tribes. Faithful, uncomplaining, unflinching, the soldiers wearing the national uniform lived for many weary years at their lonely little posts, facing unending toil and danger with quiet endurance, surrounded by the desolation of vast solitudes, and menaced by the most merciless of foes. Hunting was followed not only as a sport, but also as the only means of keeping the posts and the expeditionary trains in meat. Many of the officers became equally proficient as marksmen and hunters. The three most famous Indian fighters since the Civil War, Generals Custer, Miles, and Crook, were all keen and successful followers of the chase.

Of American big game the bison, almost always known as the buffalo, was the largest and most important to man. When the first white settlers landed in Virginia the bison ranged east of the Alleghanies almost to the sea-coast, westward to the dry deserts lying beyond the Rocky Mountains, northward to the Great Slave Lake and southward to Chihuahua. It was a beast of the forests and mountains, in the Alleghanies no less than in the Rockies; but its true home was on the prairies and the high plains. Across these it roamed, hither and thither, in herds of enormous, or incredible magnitude; herds so large that they covered the waving grass land for hundreds of square leagues, and when on the march occupied days and days in passing a given point. But the seething myriads of shaggy-maned wild cattle vanished with remarkable and melancholy rapidity before the inroads of the white hunters, and the steady march of the oncoming settlers. Now they are on the point of extinction. Two or three hundred are left in that great national game preserve, the Yellowstone Park; and it is said that others still remain in the wintry desolation of Athabasca. Elsewhere only a few individuals exist—probably considerably less than half a hundred all told—scattered in small parties in the wildest and most remote and inaccessible portions of the Rocky Mountains. A bison bull is the largest American animal. His huge bulk, his short, curved black horns, the shaggy mane clothing his great neck and shoulders, give him a look of ferocity which his conduct belies. Yet he is truly a grand and noble beast, and his loss from our prairies and forest is as keenly regretted by the lover of nature and of wild life as by the hunter.

Next to the bison in size, and much superior in height to it and to all other American game—for it is taller than the tallest horse—comes the moose, or broad-horned elk. It is strange, uncouth-looking beast, with very long legs, short thick neck, a big, ungainly head, a swollen nose, and huge shovel horns. Its home is in the cold, wet pine and spruce forests, which stretch from the sub-arctic region of Canada southward in certain places across our frontier. Two centuries ago it was found as far south as Massachusetts. It has now been exterminated from its former haunts in northern New York and Vermont, and is on the point of vanishing from northern Michigan. It is still found in northern Maine and northeastern Minnesota and in portions of northern Idaho and Washington; while along the Rockies it extends its range southward through western Montana to northwestern Wyoming, south of the Tetons. In 1884 I saw the fresh hide of one that was killed in the Bighorn Mountains.

The wapiti, or round-horned elk, like the bison, and unlike the moose, had its centre of abundance in the United States, though extending northward into Canada. Originally its range reached from ocean to ocean and it went in herds of thousands of individuals; but it has suffered more from the persecution of hunters than any other game except the bison. By the beginning of this century it had been exterminated in most localities east of the Mississippi; but a few lingered on for many years in the Alleghanies. Colonel Cecil Clay informs me that an Indian whom he knew killed one in Pennsylvania in 1869. A very few still exist here and there in northern Michigan and Minnesota, and in one or two spots on the western boundary of Nebraska and the Dakotas; but it is now properly a beast of the wooded Western mountains. It is still plentiful in western Colorado, Wyoming, and Montana, and in parts of Idaho, Washington, and Oregon. Though not as large as the moose it is the most beautiful and stately of all animals of the deer kind, and its antlers are marvels of symmetrical grandeur.

The woodland caribou is inferior to the wapiti both in size and symmetry. The tips of the many branches of its long irregular antlers are slightly palmated. Its range is the same as that of the moose, save that it does not go so far southward. Its hoofs are long and round; even larger than the long, oval hoofs of the moose, and much larger than those of the wapiti. The tracks of all three can be told apart at a glance, and can not be mistaken for the footprints of other game. Wapiti tracks, however, look much like those of yearling and two-year-old cattle, unless the ground is steep or muddy, in which case the marks of the false hoofs appear, the joints of wapiti being more flexible than those of domestic stock.

The whitetail deer is now, as it always has been, the best known and most abundant of American big game, and though its numbers have been greatly thinned it is still found in almost every State of the Union. The common blacktail or mule deer, which has likewise been sadly thinned in numbers, though once extraordinarily abundant, extends from the great plains to the Pacific; but is supplanted on the Puget Sound coast by the Columbian blacktail. The delicate, heart-shaped footprints of all three are nearly indistinguishable; when the animal is running the hoof points are of course separated. The track of the antelope is more oval, growing squarer with age. Mountain sheep leave footmarks of a squarer shape, the points of the hoof making little indentations in the soil, well apart, even when the animal is only walking; and a yearling's track is not unlike that made by a big prong-buck when striding rapidly with the toes well apart. White-goat tracks are also square, and as large as those of the sheep; but there is less indentation of the hoof points, which come nearer together.

The antelope, or prong-buck, was once found in abundance from the eastern edge of the great plains to the Pacific, but it has everywhere diminished in numbers, and has been exterminated along the eastern and western borders of its former range. The bighorn, or mountain sheep, is found in the Rocky Mountains from northern Mexico to Alaska; and in the United States from the Coast and Cascade ranges to the Bad Lands of the western edges of the Dakotas, wherever there are mountain chains or tracts of rugged hills. It was never very abundant, and, though it has become less so, it has held its own better than most game. The white goat, however, alone among our game animals, has positively increased in numbers since the advent of settlers; because white hunters rarely follow it, and the Indians who once sought its skin for robes now use blankets instead. Its true home is in Alaska and Canada, but it crosses our borders along the lines of The Rockies and Cascades, and a few small isolated colonies are found here and there southward to California and New Mexico.

The cougar and wolf, once common throughout the United States, have now completely disappeared from all save the wildest regions. The black bear holds its own better; it was never found on the great plains. The huge grisly ranges from the great plains to the Pacific. The little peccary or Mexican wild hog merely crosses our southern border.

The finest hunting ground in America was, and indeed is, the mountainous region of western Montana and northwestern Wyoming. In this high, cold land, of lofty mountains, deep forests, and open prairies, with its beautiful lakes and rapid rivers, all the species of big game mentioned above, except the peccary and Columbian blacktail, are to be found. Until 1880 they were very abundant, and they are still, with the exception of the bison, fairly plentiful. On most of the long hunting expeditions which I made away from my ranch, I went into this region.

The bulk of my hunting has been done in the cattle country, near my ranch on the Little Missouri, and in the adjoining lands round the lower Powder and Yellowstone. Until 1881 the valley of the Little Missouri was fairly thronged with game, and was absolutely unchanged in any respect from its original condition of primeval wildness. With the incoming of the stockmen all this changed, and the game was wofully slaughtered; but plenty of deer and antelope, a few sheep and bear, and an occasional elk are still left.

Since the professional hunters have vanished with the vast herds of game on which they preyed, the life of the ranchman is that which yields most chance of hunting. Life on a cattle ranch, on the great plains or among the foothills of the high mountains, has a peculiar attraction for those hardy, adventurous spirits who take most kindly to a vigorous out-of-door existence, and who are therefore most apt to care passionately for the chase of big game. The free ranchman lives in a wild, lonely country, and exactly as he breaks and tames his own horses, and guards and tends his own branded herds, so he takes the keenest enjoyment in the chase which is to him not merely the pleasantest of sports, but also a means of adding materially to his comforts, and often his only method of providing himself with fresh meat.

Hunting in the wilderness is of all pastimes the most attractive, and it is doubly so when not carried on merely as a pastime. Shooting over a private game preserve is of course in no way to be compared to it. The wilderness hunter must not only show skill in the use of the rifle and address in finding and approaching game, but he must also show the qualities of hardihood, self-reliance, and resolution needed for effectively grappling with his wild surroundings. The fact that the hunter needs the game, both for its meat and for its hide, undoubtedly adds a zest to the pursuit. Among the hunts which I have most enjoyed were those made when I was engaged in getting the winter's stock of meat for the ranch, or was keeping some party of cowboys supplied with game from day to day.

# 3

# The Big-Horn Sheep

IT HAS HAPPENED THAT I HAVE GENERALLY HUNTED BIG-horn during weather of arctic severity; so that in my mind this great sheep is inseparably associated with snow-clad, desolate wastes, ice-coated crags, and the bitter cold of a northern winter; whereas the sight of a prong-buck, the game that we usually hunt early in the season, always recalls to me the endless green of the midsummer prairies as they shimmer in the sunlight.

Yet in reality the big-horn is by no means confined to any one climatic zone. Along the interminable mountain chains of the Great Divide it ranges south to the hot, dry table-lands of middle Mexico, as well as far to the northward of the Canadian boundary, among the towering and tremendous peaks where the glaciers are fed from fields of everlasting snow. There exists no animal more hardy, nor any better fitted to grapple with the extremes of heat and cold. Droughts, scanty pasturage, or deep snows make it shift its ground, but never mere variation of temperature. The lofty mountains form its favorite abode, but it is almost equally at home in any large tract of very rough and broken ground. It is by no means an exclusively alpine animal, like the white goat. It is not only found throughout the main chains of the Rockies, as well as on the Sierras of the south and the coast ranges of western Oregon, Washington, and British Columbia, but it also exists to the east among the clusters of high hills and the stretches of barren Bad Lands that break the monotonous level of the great plains.

Throughout most of its range the big-horn is a partly migratory beast. In the summer it seeks the highest mountains, often passing above timber-line; and when the fall snows deepen it comes down to the lower spurs or foot-hills, or may even travel some distance southward. If there is a large tract of Bad Lands near the mountains, sheep may be plentiful in them throughout the severe weather, while in the summer not a single individual will be found in its winter haunts, all having then retired to the high peaks.

Sometimes big-horn wander widely for reasons unconnected with the weather: all of those in a district may suddenly leave it and perhaps not return for several years. Such is often the result of a district being settled, or being exposed to incessant hunting. After a certain number of sheep have been killed the remainder may all disappear, possibly one or two small bands only staying behind; but it is quite likely that two or three years later the bulk of the vanished host will come back again.

But where the region that they inhabit is cut off from the mountains by settled districts, or by great stretches of plain and prairie, then the sheep that dwell therein can make no such migrations. Thus they live all the year round in the Little Missouri Bad Lands; and though the different bands wander away and to and fro for scores of miles, especially in the fall,—for big-horn are far more restless than deer,—yet they do not shift their positions much on account of the season, and are often found in precisely the same places both summer and winter. They thus bear with indifference exposure to the extremes of heat and cold in a climate where the yearly variation reaches the utmost possible limit, the thermometer sometimes covering a range of a hundred and seventy degrees in the course of twelve months. There are few spots on earth much hotter than these Bad Lands during a spell of fierce summer weather, and, unlike the deer, the sheep cannot seek the shade of the dense thickets. In the glare of midday the naked angular hills yield no shelter whatever; the barren ravines between them turn into ovens beneath the brazen sun. The still, lifeless, burning air stifles those who breathe it, while the parched and heat-cracked cañon walls are intolerable to touch.

But though the mountain sheep can stand this, and in fact do so with even less protection than the deer, yet they certainly dislike it more than do the latter. If mountains are near, they go up them far sooner and far higher than the deer. On the other hand, they bear the winter blizzards much better, caring less for shelter, and keeping their strength pretty well. Ordinarily when in the Bad Lands they do not shift their ground save to get on the lee side of the cliffs, though the deep snows of course drive them from the mountains. A very heavy fall of snow, if they are high up on the hills, occasionally forces a band to enter the evergreen woods and make a regular yard, as deer do, beneath the overhanging cover-giving branches; then they subsist on the scanty browse until they can get back to pasture lands. But this is rare. Generally they stay in the open, and bid defiance to the elements; yet, like other game, they often seem to have the knack of foretelling any storm or cold spell of unusual severity and length. On the eve of such a storm they frequently retreat to some secure haven of refuge. This may be a nook or cranny in the rocks, or merely a slight hollow to leeward of a little grove of stunted pines; and there the band may have to stay without food for several days, until the storm is over. Occasionally they succumb to the deep snow; but if they have any kind of chance for their lives, this happens less often than with either deer or antelope.

The big-horn, or cimarrón sheep, as the Mexicans call it, is the sole American representative of the different kinds of mountain sheep that are found in the Old World. It is fourfold the weight of the Mediterranean moufflon. Its nearest relative, from which it is with difficulty distinguished, is the huge argali, three or four varieties—some say species—of which are to be found in the high lands of central Asia. The American and Asiatic animals seem to grade into one another as regards size; the north Asiatic argali is said to be no larger than the big-horn but the giant Himalayan sheep, or nyan, averages heavier, both in body and horns, and especially in length of legs. The horns of the argali have more outward twist. The largest big-horn of which I have ever been able to get authentic record was one killed in Montana by a ranch friend of mine, and carefully weighed and measured at the time. At the shoulder he stood just three feet eight inches; he weighed very nearly four hundred pounds; and his single unbroken horn was in girth nineteen inches, and in length along the curve forty-two. But such a ram is a giant. The largest I have myself shot I had no means of weighing: it was just after the rutting season, and he was as gaunt as a greyhound. At the shoulder he stood three feet five inches; and his horns, which were thick for their length, were in girth sixteen and a half inches, and in length thirty. The nyan of Thibet, on the other hand, stands four feet high; and exceptional rams have horns twenty-three inches round the base and upwards of fifty in length, while the average full-grown one will perhaps have them seventeen inches by thirty-eight. The nyan thus certainly stands before the big-horn, although even among full-grown animals many heads of the latter would be above the average of the former. The difference in the habits of the two animals is very marked, for according to the English sportsmen the nyan keeps exclusively to the high, open plains, or barren, gently sloping hills; whereas the big-horn, like the Old World ibex, is a beast of the crags and precipices, and though sometimes venturing into the level country, yet at the first alarm it invariably dashes for the broken ground.

Our American mountain sheep usually go in bands of from fifteen to thirty individuals, occasionally of many more; while often small parties of two or three will stay by themselves. In the winter, or sometimes not until the early spring, the old rams separate. The oldest and finest are often found entirely alone, retiring to the most inaccessible solitudes; the younger ones keep in little flocks of perhaps half a dozen or so. The main band then consists only of ewes, the yearlings, and now and then two-year-old; and this also is soon broken up, leaving merely the yearlings and the barren ewes, for about the middle of May the ewes that are heavy with young leave the rest, each by herself. Like the old rams, they now seek the most inaccessible and far-off places—high up the mountains, if possible; otherwise, in the barren and unfrequented portions of the Bad Lands, where the steep hills and abrupt valleys are twisted into a mere tangle of precipitous crests and cañons. Here the ewe makes her lying-in bed—oval in shape, like that of a prong-horn or black-tail doe, but made by pawing out, or perhaps merely wearing out, a slight hollow in the bare soil; whereas the doe crushes down with her weight the long grass of the prairie or thicket. This bed is usually made on the ledge of a cliff, on the side where there is most shelter from the prevailing winds; perhaps it is beneath a great rock or clay bowlder, with not so much as a blade of grass around, or it may be partly screened by a few wind-beaten sage-bushes. Generally only one, but sometimes two, young are brought forth at a birth. The young lamb matches his surroundings wonderfully in color, and the ewe is very careful in going to him to be sure that she is unobserved. For the first day or two the lamb trusts for his safety solely to not being seen by the beasts and birds of prey. He crouches flat down, like an antelope fawn, and it is next to impossible for human eyes to discover him save by accident. Once only I stumbled across a newly born lamb. It was about the first of June, and I found him lying by the bed of the mother as I was going along a ledge, scantily covered with sage brush, in the heart of some high, wild hills, about fifteen miles from my ranch. The little fellow was too young to show much alarm when I handled and petted him and with much difficulty persuaded him to stand up on his helplessly weak and awkward little legs. The mother was about two hundred yards distant, and was greatly frightened when I drew near her offspring; she hung about in the distance for a short time and then dashed off. However, she must have returned when I left; for two or three days later, when from curiosity I came back, the little fellow was gone.

When the young are able to clamber about for short distances almost as well as the old, then the nursing ewes and their lambs rejoin the band, some time in July. The band now keeps in the neighborhood of water and where the feed is good—comparatively good, at least, for the scanty pasturage that grows on the mountains and barren hills haunted by the sheep would hardly please more luxury-loving animals. The flocks of ewes and lambs are at this time quite easily discovered, but of course no man but a game butcher would dream of molesting them. In September the young rams begin to join them, and soon afterwards the old patriarchs likewise come down from their remote fastnesses.

The rams now fight desperately among themselves for the possession of the ewes, rushing together with a shock that would shatter their skulls were they less strong; while the battered horns, with splintered ends, bear witness to the violence of the contests. These contests are free from one danger, however; the horns do not get interlocked, and thus cause the death of both combatants. This is not only a common accident among deer and elk, but it even happens to antelope; I knew of one instance where two prong-horn bucks, who had evidently been battling for a doe, were found dead, side by side, partly eaten by the coyotes. The right horn of one and the left horn of the other had become locked together so firmly, thanks to the prong and hook at the end, that they could not be drawn apart, and the two beasts had died miserably in consequence. Each herd has some acknowledged master ram, but he may tolerate the presence of three or four others of lesser degree, together with the ewes, lambs, and yearlings that go to make up the rest of the flock; or else, if a cross old fellow, the master ram may turn out all the others, or may content himself with a little bunch of merely three or four ewes. So that even at this season several young rams may be found by themselves; or a morose old veteran, time-worn and battle-scarred, may keep entirely alone. As soon as the rutting season is over many of these exiles rejoin the band; and at this time, when the rams are of course in very poor condition, they are all apt to come down on the levels more boldly than at any other season, to get at the good grass, although even now rarely venturing very far from the hills. While thus on the edges of the plains, their natural wariness seems to increase tenfold.

But at all times their habits are very variable; for they are restless, wandering beasts, with something whimsical in their tempers, and given at times to queer freaks. If the fit seize them, and especially if they have been alarmed or annoyed, they may at any time leave their accustomed dwelling-places, or act in a manner absolutely contrary to their usual conduct. About noon one hot mid-summer day, three great rams crossed the river just below our ranch, stopping to drink, and spending some time on the sand-bars, occasionally playfully butting at each other. They trotted off before they could be stalked. To get down to the river they had to pass over a level plain half a mile wide; and once across, they went through a dense wood choked with underbrush for nearly half a mile more before again coming to the steep bluffs. On another occasion, in the rutting season, one of my cowboys encountered a mountain-ram crossing a broad, level river-bottom at midday. Occasionally a ram will join a flock of ewes, or a ewe and a yearling, in the spring. Two or three times I have known them to come boldly up to the bluffs that overlook and skirt a little frontier town, and there to stay grazing or resting for several hours; but they always made off in plenty of time to avoid the hunters who finally went after them. Once I shot one within a few hundred yards of my ranch house. I was returning home, weary and unsuccessful after a long day's tramp over hills where black-tail usually were common. When nearly home I struck into a well-beaten cattle-trail, leading down a deep, narrow ravine which cleft in two a knot of jagged hills; it was a favorite range for our horses, and so was frequently ridden over by the cowboys. On turning round a corner of the ravine, a sudden snort to one side and above me made me hastily look up, shifting my rifle from my shoulder. On my right the sheer wall of clay rose up without a break for perhaps two hundred feet or so, its thin, notched crest showing against the sky-line as sharply as if cut with a knife; and on a little jutting pinnacle was perched a mountain sheep, its four hoofs all together on a space no larger than the palms of a man's hands. It was facing me and staring down at me, so that the bullet went right into its chest, splitting its heart fairly open. Yet it did not fall forward over the cliff, but wheeled on its haunches and went along the crest at a mad, plunging gallop, finally crossing out of sight. Almost as soon as it disappeared a column of dust rose from the other side of the ridge, making me think that it had fallen some distance, striking hard on the dry clay. The guess was a good one, and when, after a long circle and some climbing, I reached the spot, I found a fine young barren ewe lying dead at the foot of a high cut bank.

But all such instances as these are wholly exceptional, and are chiefly interesting as showing that mountain sheep act more erratically and less according to rule than do most other kinds of game. They seem to have fits of restless waywardness, or even of panic curiosity; and so at times wander into unlooked-for places, or betray a sudden heedlessness of dangers against which they on ordinary occasions carefully guard. This last freak, however, is generally shown only in very wild localities or among young animals. Where hunters are scarce or almost unknown, all wild animals are very bold. I have seen deer in remote forests, and even in little-hunted localities near my ranch, so tame that they would stand looking at the hunter within fifty yards for several minutes before taking flight. Mountain sheep under similar circumstances show a lordly disregard for the human in strong contrast to the mad gallop of their more sophisticated brethren when alarmed.

In fact, much of the wariness among beasts of chase, as well as much of the courage shown by the more ferocious, depends upon the degree in which they have been harried by hunters, although much also depends upon the character of the species. European game is thus generally wilder than American; but no animal could be more difficult to approach than a Maine moose. The deer of the Adirondacks and Alleghanies are almost as wary, and in those parts of the Rockies where they have been much molested, big-horn are as shy as the chamois of the Alps, or the ibex of the Pyrenees. So the sloth bear and leopard of India are now much more vicious and dangerous to man than are the black bear and cougar of the United States, simply because of the different race of human beings by whom they are surrounded.

No animal seems to have been more changed by domestication than the sheep. The timid, helpless, fleecy idiot of the folds, the most foolish of all tame animals, has hardly a trait in common with his self-reliant wild relative who combines the horns of a sheep with the hide of a deer, whose home is in the rocks and the mountains, and who is so abundantly able to take care of himself. Wild sheep are as good mountaineers as wild goats, or as mountain antelopes, and are to the full as wary and intelligent.

A very short experience with the rifle-bearing portion of mankind changes the big-horn into a quarry whose successful chase taxes to the utmost the skill alike of still-hunter and of mountaineer. A solitary old ram seems to be ever on the watch. His favorite resting-place is a shelf or terrace-end high up on some cliff, from whence he can see far and wide over the country round about. The least sound—the rattle of a loose stone, a cough, even a heavy footfall on hard earth—attracts his attention, making him at once clamber up on some peak to try for a glimpse of the danger. His eyes catch the slightest movement. His nose is as keen as an elk's, and gives him surer warning than any other sense; the slightest taint in the air produces immediate flight in the direction away from danger. But there is one compensation, from the hunter's standpoint, for his wonderfully developed smelling powers; he lives in such very broken country that the currents of air often go over his head, so that it is at times possible to hunt him almost down wind.

A band of sheep is, if anything, even more difficult to approach than is a single ram; but, on the other hand, it is far easier to get on the track of and to find out as there are always some young members guilty of indiscretions. All of the flock are ever on the lookout. While the others are grazing there is always at least one with its head up; and occasionally a particularly watchful ewe will jump up on some bowlder, or at least stand with her fore-legs against its side, so as to get a wider view. Any unexplained sight or sound is announced to the rest of the herd by a kind of hissing snort, or sometimes by a stamp of the forefoot on the ground. If the intruder is either smelt or seen, the whole herd instantly break into the strong but not particularly swift gallop which distinguishes the species, and they go straight away from the danger towards the roughest ground that they can reach. If, however, only alarmed by a sound, or if the suspicious object is some distance off, the animals often run together into a bunch and stand gazing in its direction for a few seconds prior to making off. Among cliffs and precipices the echoes are so confusing that if the hunter keeps out of sight the herd occasionally become utterly bewildered by the firing, and, as a result, spend several fatal minutes in a futile running to and fro, uncertain what course will take them out of danger. One day my cousin, West Roosevelt, after a long and careful stalk, got close up to three sheep in a very deep and narrow ravine; and although, owing to their being almost underneath him, he at first over shot, yet all three of the startled and panic-struck animals were killed before they recovered their wits sufficiently to run out of range.

But a chance like this may not happen once in a hunter's lifetime. Of all American game, this is the one in whose pursuit the successful hunter needs to show most skill, hardihood, and resolution. On ordinary occasions a big-horn, when menaced by danger, flees beyond its reach with instant decision and headlong speed, disappearing with incredible rapidity over ground where it needs an expert cragsman to so much as follow at a walk. Its wonderful feats of climbing have, as with the chamois and ibex of the Old World, given rise to many fables, the most widespread being the belief that the rams, in plunging down precipices, alight on their horns. So the chamois was said to hang over ledges by means of its short, hooked horns, and when cornered on the edge of a sheer precipice, where there was no escape from the hunter, of its own accord to thrust its body against his outstretched knife—as we read and see pictured in the German hunting-books of two or three centuries ago, such as the quaint old "Adeliche Weidwerke."

The mountain sheep of America, when the choice is open to them, actually seem to prefer regions as wild and rugged as they are sterile. The tufts of grass between the rocks, the scanty blades that grow on the clay buttes, suffice for their wants, and the amount of climbing necessary to get at them is literally a matter of indifference to beasts whose muscles are like whip-cord and whose tendons are like steel. A big-horn is a marvelous leaper, perhaps even better when the jump is perpendicular than when it is horizontal. His poise is perfect; his eye and foot work together with unerring accuracy. One will unhesitatingly bound or drop a dozen feet on to a little rock pinnacle where there is scarce a hand's breadth on which to stand. The presence of the tiniest cracks in the otherwise smooth surface of a sheer rock wall enables a mountain sheep to go up it with ease. The proud, lordly bearing of an old ram makes him look exactly what he is, one of the noblest of game animals; his port is the same whether at rest or in motion. Except when very badly frightened, his movements are all made with a certain self-confident absence of hurry, as if he were conscious of a vast reserve power of strength and activity on which to draw at need. As a mountaineer he is the embodiment of elastic, sinewy strength and self-command rather than of mere nervous agility. He hardly ever makes a mistake, even when rushing at speed over the slippery, ice-coated crags in winter.

The most difficult of all climbing is to go over rocks when the ice has filled up all the chinks and crannies and the flat slabs are glassy in their hard smoothness. A black-tail buck is no mean climber; yet under such circumstances I have seen one lose his footing and tumble head over heels, scraping great handfuls of hair off his hide; but I have never known a big-horn to make a misstep. This is undoubtedly largely owing to the difference between the two animals in the structure of their feet. A sheep's hoof is an elastic pad, only the rims and the toe-points being hard, and it thus gets a good grip on the slightest projection, or on any little roughness in the rock. The tracks are very different from deer tracks, being nearly square in form, instead of heart-shaped, the prints of the toes rather deep and wide apart, even when the animal has been walking.

A band of sheep will often seem to court certain death by plunging off the brink of what looks like a perpendicular cliff, where there is not a ledge or a crack yielding foot-hold. In such cases, if the cliff is high, it will be found on examination that it is not quite perpendicular, and that the sheep, in making the fearful descent, from time to time touch or strike the cliff with their hoofs thus going down in long bounds, keeping their poise all the time. The final bound is often made almost head first, as if they were diving.

Narrow ledges, overlooking an abyss the fathomless depths of which would make even a trained cragsman giddy, are very favorite resorts. So are the crests of the ridges themselves. If in any patch of Bad Lands there is an unusually high chain of steep, bare clay buttes, mountain sheep are sure to select their tops as a regular parade-ground. After a rain the clay takes their hoof-prints as clearly as if it were sealing-wax, and all along the top of the crest they beat out a regular walk from one end to the other, with occasional little side-paths leading out to some overhanging shoulder or jutting spur, from whence there is a good view of the surrounding country.

Generally the band is led by a ewe; but in a case of immediate and pressing danger the ram assumes the headship. Aside from man, mountain sheep have fewer foes than most other game. Bears are too clumsy to catch them; and lynx and fox, inveterate enemies of fawns, rarely get up to the high, breezy nurseries of the young lambs. Wolves and cougars, however, harass them greatly. A wolf will not attack an old ram if he can help it, but sneaks after the ewes and lambs, waiting until they get on somewhat level ground, and then running one down by sheer speed before it can take refuge among the secure fastnesses of the precipices.

The cougar relies on stealth, not on speed, and gets his game either by fair stalking or else by lying in wait. Sometimes he can creep up to a band while they are taking their siesta; but generally they keep too sharp a lookout and he has to approach them while they are feeding, or when they have come down to drink. Some fifteen miles from my ranch is a tract of very rough country, the sides of the hills falling off into precipices or into dark, cedar-clad gorges. This was a favorite resort of mountain sheep; but one spring a couple of cougars took up their abode in the neighborhood, and soon killed several of the sheep and drove the others away. Judging by the tracks and by the position of the carcasses, they must have done the killing in the morning and evening, creeping up to the doomed animals as they fed on the lower slopes, or lurking round the spring-holes and little alkali pools where they drank. The great war eagle is one of the worst enemies of the young lambs.

In the rutting season a ram will make a good fight if he has any chance at all, and at that time is very bold and pugnacious. If followed by a dog he will frequently decline to run, turning to bay at once. One hunter whom I knew killed several in this way by the aid of a collie. Of course it cannot be done when once the sheep have begun to realize that the dog is merely an ally of the man, for they then look out for the latter.

Sheep are easily tamed, if taken young, and make amusing pets. A friend in Helena, Montana, once owned a tame ram. When young he was a great favorite. He was an inquisitive, mischievous creature, of marvelous activity. It was impossible to keep him out of the garden. A single hop would carry him over the high fence; if an inmate of the house came to the rescue, another hop carried the intruder once more into outside safety, and a third took him back again the second the rescuer had turned around. Whenever he got the chance he would pull down the clothes that had been hung up to dry. When he could get inside the house he was fond of walking on the mantel-piece. He was the terror of the Chinese cook, whom he soon discovered to be afraid of him, and would lie in wait outside the kitchen door so as to butt him when he appeared. This was at first done in mere playfulness; but as he grew older he became morose and quarrelsome, and had to be disposed of.

It is impossible to hunt big-horn successfully without some

knowledge of their habits. They go down to drink in the very late evening or sometimes in the gray of the morning; when the moon is full they may not go to the water until long after nightfall. Generally they drink later than any other game; but all game vary their habits now and then in this regard. The prong-buck, though diurnal, sometimes comes to a watering-hole during the night; and I have once or twice seen both deer and sheep drinking at midday.

In ordinary weather they begin to feed early in the morning, and when the sun has risen some little distance above the horizon they start to graze their way slowly up to the high spur or ridge crest where they intend to lie during the day. Here they stay until well on in the afternoon, and then again descend to their feeding-grounds on the lower slopes. In very cold weather, however, they are apt to be found grazing at midday. A raging snow blizzard may keep them lying close under cover for three days at a time: they naturally get ravenous, and when there is a lull, or especially if it is succeeded by a short spell of good weather, they come hastily out to feed, no matter what the time of day may be.

As with almost all game except antelope, they can be best hunted in the morning and evening; but, unlike deer, they can also be followed throughout the day, for whereas elk, black-tail, and white-tail have then all alike retired to the thickets, the big-horn take their noontide rest lying out in plain view. If the hunter means to catch them feeding he should make a very early start. A good pair of field-glasses is of great service, for the two essential requisites to success are the capacity to take long walks over rough ground and painstaking care in scanning the country far and wide, so as to see the game before it sees the hunter. There is then a chance to stalk up close, the broken ground frequently yielding good cover.

Often it may be necessary to lie for hours carefully concealed, watching a flock that is in an unfavorable position, and waiting until it shifts its ground. This is not very comfortable on a cold day in November or December, the months in which I have usually hunted big-horn, devoting the early fall to the chase of elk and deer. But it is often the only way to secure success: patience and perseverance are two of the still-hunter's cardinal virtues. Personally I have always owed whatever success I have had to dogged perseverance and patient persistence, and on a lamentably large number of occasions have had to draw heavily on these qualities to make good a lack of skill, sometimes with the rifle, sometimes in mountaineering. Among many hunting trips I can recall not a few where willingness to lie still two or three hours under trying circumstances in the end got me the game; and one such instance may serve as a sample of the rest.

I was staying at the line camp of two of my cowboys, a small dug-out in the side of a butte that marked the edge of the Bad Lands, the rolling prairie coming up to its base. The quarters were cramped for three men, an entire side of the little hut being filled by the two bunks in which we slept,—I in the upper, my two companions in the lower,—while the fireplace occupied one end, the mess-box served as a table, and the earth-covered roof of logs was so low that we could hardly stand upright. Window there was none; but it was snug, and for a line camp, clean. There was plenty of fire-wood, and, for a wonder, the chimney did not smoke; so we were comfortable enough. The butte itself served for three out of the four walls. No other building is so warm as a dug-out, and in the terrible winter weather of Dakota and Montana warmth is the one thing for which all else must be sacrificed.[1]

In such high latitudes the December sun rises late. Long before daybreak we had finished our breakfast of bread, beans, and coffee. The two cowboys had saddled their shaggy ponies—who had spent the night in the rough log stable—and had ridden off in opposite directions along their lonely beat, muffled in their wolf-skin overcoats and heavy shaps; while I strode off on foot towards the high hills that lay riverward, my rifle on my shoulder and my fur cap pulled down well over my ears.

The cold was biting, for even at noon the sun had not power to thaw the frozen ground. But there was very little snow; just enough to powder the hills and to lie in patches in the hollows. I walked rapidly up a long coulée, then climbed up a steep rounded hill that followed the divide back into the heart of the Bad Lands. By the time I was on my chosen hunting-grounds the sun had topped the horizon behind me, and his level rays lit up the peaks and crests.

The next hour was spent in hard climbing and incessant watchfulness. The hills lay in isolated masses. I clambered painfully up their slippery sides, creeping along the narrow icy ledges that ran across the faces of the cliffs, and cautiously working my way over the smooth shoulders. From behind every ridge and spur I carefully examined the opposite hill-sides, using the field-glasses if there was scope for them. Sheep, standing still or lying down, are often very hard to see, their coats assimilating curiously with the neutral-tinted cliffs and bowlders; but against snow they of course stand out much more distinctly.

At last, as I lay peeping over the ragged crest of a clay butte, I made out a small dark object half way up a steep slope some six hundred yards down the valley; and another look showed me that it was a ram feeding leisurely up the hill-side. The wind was good for a direct approach. I got off the butte by carefully letting myself down from one little ledge or niche to another, and started along the valley towards the ram, only to find my way barred by a deep chasm whose straight, ice-coated sides yawned too far apart to permit of any attempt at crossing. There was no help for it but laboriously to retrace my steps and make my way round its head with what speed I could. This I did, the work making me thoroughly warm for the first time that morning. Once across the walking was better, and I went down the valley-side at a good pace, until I came to a shoulder some two hundred yards from where I had seen the sheep. I was a good deal higher than where he had stood; but in the time I had been out of sight of him he must have gone up the hill quite a distance, for when I looked round the shoulder I saw him about as far off as I expected, but above instead of below me. Slow though my movements had been when I cautiously looked round the edge, they had not escaped his quick eye; for when I made him out he was standing motionless, gazing in my direction. Before I could raise my rifle he gave a great jump sideways and galloped off, disappearing instantly behind a huge mass of detached sandstone, and I never saw him again.

A little chagrined at my fruitless stalk I plodded on, doing much hard climbing but seeing no signs of game until nearly midday. Then in the snow at the head of a coulée I came across the tracks of a band evidently made that morning while returning from the feeding-grounds. I followed them until I became convinced that the animals had gone to a great tableland or plateau that I could see a good way ahead; then, as the wind was behind me, I struck off to one side, made a circle through some very rough country, and clambered out along the knife-like crests of a line of high hills separated from the plateau by a broad valley. Every hundred paces or so I would stop and examine the country far and near with the glasses; often I had to crawl on all-fours to avoid appearing against the sky-line on the ridge.

At last I caught sight of the band. There were some fifteen or twenty of them, and they were lying at the point of a spur that was thrust out from the plateau, nearly opposite to me and half a mile off. They were in a position which it was impossible to approach within six hundred yards without being observed, for they could see over the level plateau behind them, and from the brink of the lofty cliff on which they lay they looked up, down, and across the wild, deep valley beneath.

With the glasses I could make out that there was no good head among them; but I was out for meat rather than for sport. They were very watchful, ever on the lookout; and as the afternoon wore on one of the more restless would now and then get up, walk off a few steps, or stand gazing intently into the far distance. There was nothing for me to do except to wait until they grew hungry and shifted their position to some place which there was a chance of my approaching unseen. So for the three hours I lay on the iron ground, under the lee of a bowlder that but partly shielded me from the wind, munching the strip of jerked venison I had carried in my pocket, and peeping at the sheep through a tuft of tall, coarse grass that grew on top of the ridge.

At last, when it wanted but little more than an hour of sunset, the sheep all got on their legs, one after another, and, led by an old ewe, began to descend into the valley. They went down the cliff by a sort of break or slide, hopping dexterously from rock to rock. On coming to the steep slope at its foot they struck into a trot, which merged into a fast gallop as they got nearly down. I feared that they would stop before coming to the cañon at the bottom of the valley; but they did not, crossing it without hesitation, for all its sheer-sided and slippery depth, and continuing their course towards the end of the chain of hills on which I was, where they halted to graze, after going up nearly to the top. It was excellent ground for a stalk. The ridge went down to the left in the steep, grassy slopes on which they were feeding, while on the right it broke abruptly off into a precipice, with a narrow ledge high up along its face.

This ledge made the approach an easy one. The only difficult places were those where the ledge was interrupted, and I had either cautiously to make my way along the face of the cliff,—a very unpleasant task, as the slight hollows or knobs which served me as foot-holds were slippery with ice, the risk of a fall being thus enormously increased,—or else was forced to go to the top, and, sprawling flat on the smooth slope, drag myself along just to one side of the ridge. I had marked the position of the game by a dwarfed cedar that grew in a crevice on the very crest. It gave excellent cover, and on reaching it and peering out through the branches, I saw the sheep scattered out only some sixty yards below me, and, choosing out a fine young ram, I fired, breaking both shoulders. They all rushed together, and then without an instant's pause raced madly down the hill-side, neither of the two bullets that I sent after them taking effect. I had no time to lose; so I dressed the ram hastily, tilted him up so that the blood would run out, and left him to be called for with the pony next day. Then I made the best use of the waning light to get to a long divide, furrowed by many buffalo trails, which I knew I could follow even when it grew dark, and which came out on the prairie not very far to one side of the line camp.

The day on which I was lucky enough to shoot my largest and finest ram was memorable in more ways than one. The shot was one of the best I ever made,—albeit the element of chance doubtless entered into it far more largely than the element of skill,—and in coming home from the hunt I got quite badly frozen.

The day before we had come back from a week's trip after deer; for we were laying in the winter stock of meat. We had been camped far down the river, and had intended to take two days on the return trip, as the wagon was rather heavily loaded, for we had killed eight deer. The morning we broke camp was so mild that I did not put on my heaviest winter clothing, starting off in the same that I had worn during the past few days' still-hunting among the hills. Before we had been gone an hour, however, the sky grew overcast and the wind began to blow from the north with constantly increasing vigor. The sky grew steadily more gloomy and lowering, the gusts came ever harder and harder, and by noon the winter day had darkened and a furious gale was driving against us. The blasts almost swept me from my saddle and the teamster from his seat, while we were glad to wrap ourselves in our huge fur coats to keep out the growing cold. Soon after midday the wagon suddenly broke down while we were yet in mid-prairie. It was evident that we were on the eve of a furious snow-blizzard, which might last a few hours, or else, perhaps, as many days. We were miles from any shelter that would permit us to light a fire in the face of such a storm; so we left the wagon as it was, hastily unharnessed the team horses, and, with the driver riding one and leading the other, struck off homeward at a steady gallop. Once fairly caught by the blizzard in a country that we only partly knew, it would have been hopeless to do more than to try for some ravine in which to cower till it was over; so we pushed our horses to their utmost pace. Our object was to reach the head coulées of a creek leading down to the river but a few miles from the ranch. Could we get into these before the snow struck us we felt we would be all right, for we could then find our way home, even in pitch-darkness, with the wind in the quarter from which it was coming. So, with the storm on our backs, we rode at full speed through the gathering gloom, across the desolate reaches of prairie. The tough little horses, instead of faltering, went stronger mile by mile. At last the weird rows of hills loomed vaguely up in our front, and we plunged into the deep ravines for which we had been heading just as the whirling white wreaths struck us—not the soft, feathery flakes of a seaboard snow-storm, but fine ice-dust, driven level by the wind, choking us, blinding our eyes, and cutting our faces if we turned toward it. The roar of the blizzard drowned our voices when we were but six feet apart: had it not been on our backs we could not have gone a hundred yards, for we could no more face it than we could face a frozen sand-blast. In an instant the strange, wild outlines of the high buttes between which we were riding were shrouded from our sight. We had to grope our way through a kind of shimmering dusk; and when once or twice we were obliged by some impassable cliff or cañon to retrace our steps, it was all that we could do to urge the horses even a few paces against the wind-blown snow-grains which stung like steel filings. But this extreme violence only lasted about four hours. The moon was full, and its beams struggled through scudding clouds and snow-drift, so that we reached the ranch without difficulty, and when we got there the wind had already begun to lull. The snow still fell thick and fast; but before we went to bed this also showed signs of stopping. Accordingly we determined that we would leave the wagon where it was for a day or two, and start early next morning for a range of high hills some ten miles off, much haunted by sheep; for we did not wish to let pass the chance of tracking the game offered by the first good snow of the season.

Next morning we started by starlight. The snow lay several inches deep on the ground; the whole land was a dazzling white. It was very cold. Within the ranch everything was frozen solid in spite of the thick log walls; but the air was so still and clear that we did not realize how low the temperature was. Accordingly, as the fresh horse I had to take was young and wild, I did not attempt to wear my fur coat. I soon felt my mistake. The windless cold ate into my marrow; and when, shortly after the cloudless winter sunrise, we reached our hunting-grounds and picketed out the horses, I was already slightly frost-bitten. But the toil of hunting over the snow-covered crags soon made me warm.

All day we walked and climbed through a white wonderland. On every side the snowy hills, piled one on another, stretched away, chain after chain, as far as sight could reach. The stern and iron-bound land had been changed to a frozen sea of billowy, glittering peaks and ridges. At last, late in the afternoon, three great big-horn suddenly sprang up to our right and crossed the table-land in front of and below us at a strong, stretching gallop. The lengthening sunbeams glinted on their mighty horns; their great supple brown bodies were thrown out in bold relief against the white landscape; as they plowed with long strides through the powdery snow, their hoofs tossed it up in masses of white spray. On the left of the plateau was a ridge, and as they went up this I twice fired at the leading ram, my bullets striking under him. On the summit he stopped and stood for a moment to looking back three hundred and fifty yards off,[2] and my third shot went fairly through his lungs. He ran over the hill as if unharmed, but lay down a couple of hundred yards on, and was dead when we reached him.

It was after nightfall when we got back to the horses, and we rode home by moonlight. To gallop in such weather insures freezing; so the ponies shambled along at a single foot-trot, their dark bodies white with hoar-frost, and the long icicles hanging from their lips. The cold had increased steadily; the spirit thermometer at the ranch showed 26° Fahrenheit below zero. We had worked all day without food or rest, and were very tired. On the ride home I got benumbed before I knew it and froze my face, one foot, and both knees. Even my companion, who had a great-coat, froze his nose and cheeks. Never was a sight more welcome than the gleam of the fire-lit ranch windows to us that night. But the great ram's head was a trophy that paid for all.

# 4

# In the Louisiana Canebrakes

IN OCTOBER, 1907, I SPENT A FORTNIGHT IN THE CANEBRAKES of northern Louisiana, my hosts being Messrs. John M. Parker and John A. McIlhenny. Surgeon-General Rixey, of the United States Navy, and Doctor Alexander Lambert were with me. I was especially anxious to kill a bear in these canebrakes after the fashion of the old Southern planters, who for a century past have followed the bear with horse and hound and horn in Louisiana, Mississippi, and Arkansas.

Our first camp was on Tensas Bayou. This is in the heart of the great alluvial bottom-land created during the countless ages through which the mighty Mississippi has poured out of the heart of the continent. It is in the black belt of the South, in which the negroes outnumber the whites four or five to one, the disproportion in the region in which I was actually hunting being far greater. There is no richer soil in all the earth; and when, as will soon be the case, the chances of disaster from flood are over, I believe the whole land will be cultivated and densely peopled. At present the possibility of such flood is a terrible deterrent to settlement, for when the Father of Waters breaks his boundaries he turns the country for a breadth of eighty miles into one broad river, the plantations throughout all this vast extent being from five to twenty feet under water. Cotton is the staple industry, corn also being grown, while there are a few rice-fields and occasional small patches of sugar-cane. The plantations are for the most part of large size and tilled by negro tenants for the white owners. Conditions are still in some respects like those of the pioneer days. The magnificent forest growth which covers the land is of little value because of the difficulty in getting the trees to market, and the land is actually worth more after the timber has been removed than before. In consequence, the larger trees are often killed by girdling, where the work of felling them would entail disproportionate cost and labor. At dusk, with the sunset glimmering in the west, or in the brilliant moonlight when the moon is full, the cotton-fields have a strange spectral look, with the dead trees raising aloft their naked branches. The cotton-fields themselves, when the bolls burst open, seem almost as if whitened by snow; and the red and white flowers, interspersed among the burst-open pods, make the whole field beautiful. The rambling one-story houses, surrounded by outbuildings, have a picturesqueness all their own; their very looks betoken the lavish, whole-hearted, generous hospitality of the planters who dwell therein.

Beyond the end of cultivation towers the great forest. Wherever the water stands in pools, and by the edges of the lakes and bayous, the giant cypress loom aloft, rivalled in size by some of the red gums and white oaks. In stature, in towering majesty, they are unsurpassed by any trees of our Eastern forests; lordlier kings of the green-leaved world are not to be found until we reach the sequoias and redwoods of the Sierras. Among them grow many other trees—hackberry, thorn, honey-locust, tupelo, pecan, and ash. In the cypress sloughs the singular knees of the trees stand two or three feet above the black ooze. Palmettos grow thickly in places. The canebrakes stretch along the slight rises of the ground, often extending for miles, forming one of the most striking and interesting features of the country. They choke out other growths, the feathery, graceful canes standing in ranks, tall, slender, serried, each but a few inches from his brother, and springing to a height of fifteen or twenty feet. They look like bamboos; they are well-nigh impenetrable to a man on horseback; even on foot they make difficult walking unless free use is made of the heavy bush-knife. It is impossible to see though them for more than fifteen or twenty paces, and often for not half that distance. Bears make their lairs in them, and they are the refuge for hunted things. Outside of them, in the swamp, bushes of many kinds grow thick among the tall trees, and vines and creepers climb the trunks and hang in trailing festoons from the branches. Here, likewise, the bush-knife is in constant play, as the skilled horsemen thread their way, often at a gallop, in and out among the great tree-trunks, and through the dense, tangled, thorny undergrowth.

In the lakes and larger bayous we saw alligators and garfish; and monstrous snapping turtles, fearsome brutes of the slime, as heavy as a man, and with huge horny beaks that with a single snap could take off a man's hand or foot. One of the planters with us had lost part of his hand by the bite of an alligator; and had seen a companion seized by the foot by a huge garfish from which he was rescued with the utmost difficulty by his fellow swimmers. There were black bass in the waters, too, and they gave us many a good meal. Thick-bodied water-moccasins, foul and dangerous, kept near the water; and farther back in the swamp we found and killed rattlesnakes and copperheads.

Coon and possum were very plentiful, and in the streams there were minks and a few otters. Black squirrels barked in the tops of the tall trees or descended to the ground to gather nuts or gnaw the shed deer-antlers—the latter a habit they shared with woodrats. To me the most interesting of the smaller animals, however, were the swamp-rabbits, which are thoroughly amphibious in their habitats, not only swimming but diving, and taking to the water almost as freely as if they were muskrats. They lived in the depths of the woods and beside the lonely bayous.

Birds were plentiful. Mocking-birds abounded in the clearings, where, among many sparrows of more common kind, I saw the painted finch, the gaudily colored brother of our little indigo-bunting, though at this season his plumage was faded and dim. In the thick woods where we hunted there were many cardinal-birds and winter wrens, both in full song. Thrashers were even more common; but so cautious that it was rather difficult to see them, in spite of their incessant clucking and calling and their occasional bursts of song. There were crowds of warblers and vireos of many different kinds, evidently migrants from the North, and generally silent. The most characteristic birds, however, were the woodpeckers, of which there were seven or eight species, the commonest around our camp being the handsome red-bellied, the brother of the redhead which we saw in the clearings. The most notable birds and those which most interested me were the great ivory-billed woodpeckers. Of these I saw three, all of them in groves of giant cypress; their brilliant white bills contrasted finely with the black of their general plumage. They were noisy but wary, and they seemed to me to set off the wildness of the swamp as much as any of the beasts of the chase. Among the birds of prey the commonest were the barred owls, which I have never elsewhere found so plentiful. Their hooting and yelling were heard all around us through the night, and once one of them hooted at intervals for several minutes at midday. One of these owls had caught and was devouring a snake in the late afternoon, while it was still daylight. In the dark nights and still mornings and evenings their cries seemed strange and unearthly, the long hoots varied by screeches, and by all kinds of uncanny noises.

At our first camp our tents were pitched by the bayou. For four days the weather was hot, with steaming rains; after that it grew cool and clear. Huge biting flies, bigger than bees, attacked our horses, but the insect plagues, so veritable a scourge in this country during the months of warm weather, had well-nigh vanished in the first few weeks of the fall.

The morning after we reached camp we were joined by Ben Lilley, the hunter, a spare, full-bearded man, with mild, gentle, blue eyes and a frame of steel and whip-cord. I never met any other man so indifferent to fatigue and hardship. He equalled Cooper's Deer-slayer in woodcraft, in hardihood, in simplicity—and also in loquacity. The morning he joined us in camp, he had come on foot through the thick woods, followed by his two dogs, and had neither eaten nor drunk for twenty-four hours; for he did not like to drink the swamp water. It had rained hard throughout the night and he had no shelter, no rubber coat, nothing but the clothes he was wearing, and the ground was too wet for him to lie on; so he perched in a crooked tree in the beating rain, much as if he had been a wild turkey. But he was not in the least tired when he struck camp; and, though he slept an hour after breakfast, it was chiefly because had nothing else to do, inasmuch as it was Sunday, on which day he never hunted nor labored. He could run through the woods like a buck, was far more enduring, and quite as indifferent to weather, though he was over fifty years old. He had trapped and hunted throughout almost all the half-century of his life, and on trail of game he was as sure as his own hounds. His observations on wild creatures were singularly close and accurate. He was particularly fond of the chase of the bear, which he followed by himself, with one or two dogs; often he would be on the trail of his quarry for days at a time, lying down to sleep wherever night overtook him; and he had killed over a hundred and twenty bears.

Late in the evening of the same day we were joined by two gentlemen, to whom we owed the success of our hunt. They were Messrs. Clive and Harley Metcalf, planters from Mississippi, men in the prime of life, thorough woodsmen and hunters, skilled marksmen, and utterly fearless horsemen. For a quarter of a century they had hunted bear and deer with horse and hound, and were masters of the art. They brought with them their pack of bearhounds, only one, however, being a thoroughly stanch and seasoned veteran. The pack was under the immediate control of a negro hunter, Holt Collier, in his own way as remarkable a character as Ben Lilley. He was a man of sixty and could neither read nor write, but he had all the dignity of an African chief, and for half a century he had been a bear-hunter, having killed or assisted in killing over three thousand bears. He had been born a slave on the Hinds plantation, his father, an old man when he was born, having been the body-servant and cook of "old General Hinds," as he called him, when the latter fought under Jackson at New Orleans. When ten years old Holt had been taken on the horse behind his young master, the Hinds of that day, on a bear-hunt, when he killed his first bear. In the Civil War he had not only followed his master to battle as his body-servant but had acted under him as sharpshooter against the Union soldiers. After the war he continued to stay with his master until the latter died, and had then been adopted by the Metcalfs; and he felt that he had brought them up, and treated them with that mixture of affection and grumbling respect which an old nurse shows toward the lad who has ceased being a child. The two Metcalfs and Holt understood one another thoroughly, and understood their hounds and the game their hounds followed almost as thoroughly.

They had killed many deer and wildcat, and now and then a panther; but their favorite game was the black bear, which, until within a very few years, was extraordinarily plentiful in the swamps and canebrakes on both sides of the lower Mississippi, and which is still found here and there, although in greatly diminished numbers. In Louisiana and Mississippi the bears go into their dens toward the end of January, usually in hollow trees, often very high up in living trees, but often also in great logs that lie rotting on the ground. They come forth toward the end of April, the cubs having been born in the interval. At this time the bears are nearly as fat, so my informants said, as when they enter their dens in January; but they lose their fat very rapidly. On first coming out in the spring they usually eat ash buds and the tender young cane called mutton-cane, and at that season they generally refuse to eat the acorns even when they are plentiful. According to my informants it is at this season that they are most apt to take to killing stock, almost always the hogs which run wild or semiwild in the woods. They are very individual in their habits, however; many of them never touch stock, while others, usually old he bears, may kill numbers of hogs; in one case an old he bear began this hog-killing just as soon as he left his den. In the summer months they find but little to eat, and it is at this season that they are most industrious in hunting for grubs, insects, frogs, and small mammals. In some neighborhoods they do not eat fish, while in other places, perhaps not far away, they not only greedily eat dead fish, but will themselves kill fish if they can find them in shallow pools left by the receding waters. As soon as the mast is on the ground they begin to feed upon it, and when the acorns and pecans are plentiful they eat nothing else, though at first berries of all kinds and grapes are eaten also. When in November they have begun only to eat the acorns they put on fat as no other wild animal does, and by the end of December a full-grown bear may weigh at least twice as much as it does in August, the difference being as great as between a very fat and a lean hog. Old he bears which in August weigh three hundred pounds and upward will toward the end of December weigh six hundred pounds, and even more in exceptional cases.

Bears vary greatly in their habits in different localities, in addition to the individual variation among those of the same neighborhood. Around Avery Island, John McIlhenny's plantation, the bears only appear from June to November; there they never kill hogs, but feed at first on corn and then on sugar-cane, doing immense damage in the fields, quite as much as hogs would do. But when we were on the Tensas we visited a family of settlers who lived right in the midst of the forest ten miles from any neighbors; and although bears were plentiful around them they never molested their corn-fields—in which the coons, however, did great damage.

A big bear is cunning, and is a dangerous fighter to the dogs. It is only in exceptional cases, however, that these black bears, even when wounded and at bay, are dangerous to men, in spite of their formidable strength. Each of the hunters with whom I was camped had been charged by one or two among the scores or hundreds of bears he had slain, but no one of them had ever been injured, although they knew other men who had been injured. Their immunity was due to their own skill and coolness; for when the dogs were around the bear the hunter invariably ran close in so as to kill the bear at once and save the pack. Each of the Metcalfs had on one occasion killed a large bear with a knife, when the hounds had seized it and the man dared not fire for fear of shooting one of them. They had in their younger days hunted with a General Hamberlin, a Mississippi planter whom they well knew, who was then already an old man. He was passionately addicted to the chase of the bear, not only because of the sport it afforded, but also in a certain way as a matter of vengeance; for his father, also a keen bear-hunter, had been killed by a bear. It was an old he, which he had wounded and which had been bayed by the dogs; it attacked him, throwing him down and biting him so severely that he died a couple of days later. This was in 1847. Mr. W. H. Lambeth sends the following account of the fatal encounter:

"I send you an extract from the 'Brother Johnathan,' published in New York in 1847:

"'Dr. Monroe Hamberlin, Robert Wilson, Joe Brazeil, and others left Satartia, Miss., and in going up Big Sunflower River, met Mr. Leiser and his party of hunters returning to Vicksburg. Mr. Leiser told Dr. Hamberlin that he saw the largest bear track at the big Mound on Lake George that he ever saw, and was afraid to tackle him. Dr. Hamberlin said, "I never saw one that I was afraid to tackle." Dr. Hamberlin landed his skiff at the Mound and his dogs soon bayed the bear. Dr. Hamberlin fired and the ball glanced on the bear's head. The bear caught him by the right thigh and tore all the flesh off. He drew his knife and the bear crushed his right arm. He cheered the dogs and they pulled the bear off. The bear whipped the dogs and attacked him the third time, biting him in the hollow back of his neck. Mr. Wilson came up and shot the bear dead on Dr. Hamberlin. The party returned to Satartia, but Dr. Hamberlin told them to put the bear in the skiff, that he would not leave without his antagonist. The bear weighed six hundred and forty pounds.'

"Dr. Hamberlin lived three days. I knew all the parties. His son John and myself hunted with them in 1843 and 1844, when we were too small to carry a gun."

A large bear is not afraid of dogs, and an old he, or a she with cubs, is always on the lookout for a chance to catch and kill any dog that comes near enough. While lean and in good running condition it is not an easy matter to bring a bear to bay; but as they grow fat they become steadily less able to run, and the young ones, and even occasionally a full-grown she, will then readily tree. If a man is not near by, a big bear that has become tired will treat the pack with whimsical indifference. The Metcalfs recounted to me how they had once seen a bear, which had been chased quite a time, evidently make up its mind that it needed a rest and could afford to take it without much regard for the hounds. The bear accordingly selected a small opening and lay flat on its back with its nose and all its four legs extended. The dogs surrounded it in frantic excitement, barking and baying, and gradually coming in a ring very close up. The bear was watching, however, and suddenly sat up with a jerk, frightening the dogs nearly into fits. Half of them turned back-somersaults in their panic, and all promptly gave the bear ample room. The bear having looked about, lay flat on its back again, and the pack gradually regaining courage once more closed in. At first the bear, which was evidently reluctant to arise, kept them at a distance by now and then thrusting an unexpected paw toward them; and when they became too bold it sat up with a jump and once more put them all to flight.

For several days we hunted perseveringly around this camp on the Tensas Bayou, but without success. Deer abounded, but we could find no bear; and of the deer we killed only what we actually needed for use in camp. I killed one myself by a good shot, in which, however, I fear that the element of luck played a considerable part. We had started as usual by sunrise, to be gone all day; for we never counted upon returning to camp before sunset. For an hour or two we threaded our way, first along an indistinct trail, and then on an old disused road, the hardy woods horses keeping on a running walk without much regard to the difficulties of the ground. The disused road lay right across a great canebrake, and while some of the party went around the cane with the dogs, the rest of us strung out along the road so as to get a shot at any bear that might come across it. I was following Harley Metcalf, with John McIlhenny and Doctor Rixey behind on the way to their posts, when we heard in the far-off distance two of the younger hounds, evidently on the trail of a deer. Almost immediately afterward a crash in the bushes at our right hand and behind us made me turn around, and I saw a deer running across the few feet of open space; and as I leaped from my horse it disappeared in the cane. I am a rather deliberate shot, and under any circumstances a rifle is not the best weapon for snap-shooting, while there is no kind of shooting more difficult than on running game in a canebrake. Luck favored me in this instance, however, for there was a spot a little ahead of where the deer entered in which the cane was thinner, and I kept my rifle on its indistinct, shadowy outline until it reached this spot; it then ran quartering away from me, which made my shot much easier, although I could only catch its general outline through the cane. But the 45–70 which I was using is a powerful gun and shoots right through cane or bushes; and as soon as I pulled the trigger the deer, with a bleat, turned a tremendous somersault and was dead when we reached it. I was not a little pleased that my bullet should have sped so true when I was making my first shot in company with my hard-riding, straight-shooting planter friends.

But no bear were to be found. We waited long hours on likely stands. We rode around the canebrakes through the swampy jungle, or threaded our way across them on trails cut by the heavy wood knives of my companions; but we found nothing. Until the trails were cut the canebrakes were impenetrable to a horse and were difficult enough to a man on foot. On going through them it seemed as if we must be in the tropics; the silence, the stillness, the heat, and the obscurity, all combining to give a certain eeriness to the task, as we chopped our winding way slowly through the dense mass of close-growing, feather-fronded stalks. Each of the hunters prided himself on his skill with the horn, which was an essential adjunct of the hunt, used both to summon and control the hounds, and for signalling among the hunters themselves. The tones of many of the horns were full and musical; and it was pleasant to hear them as they wailed to one another, backward and forward, across the great stretches of lonely swamp and forest.

A few days convinced us that it was a waste of time to stay longer where we were. Accordingly, early one morning we hunters started for a new camp fifteen or twenty miles to the southward, on Bear Lake. We took the hounds with us, and each man carried what he chose or could in his saddle-pockets, while his slicker was on his horse's back behind him. Otherwise we took absolutely nothing in the way of supplies, and the negroes with the tents and camp equipage were three days before they overtook us. On our way down we were joined by Major Amacker and doctor Miller, with a small pack of cathounds. These were good deer-dogs and they ran down and killed on the ground a good-sized bobcat–a wildcat, as it is called in the South. It was a male and weighed twenty-three and a half pounds. It had just killed and eaten a large rabbit. The stomachs of the deer we killed, by the way, contained acorns and leaves.

Our new camp was beautifully situated on the bold, steep bank of Bear Lake–a tranquil stretch of water, part of an old river-bed, a couple of hundred yards broad, with a winding length of several miles. Giant cypress grew at the edge of the water, the singular cypress knees rising in every direction round about, while at the bottoms of the trunks themselves were often cavernous hollows opening beneath the surface of the water, some of them serving as dens for alligators. There was a waxing moon, so that the nights were as beautiful as the days.

From our new camp we hunted as steadily as from the old. We saw bear sign, but not much of it, and only one or two fresh tracks. One day the hounds jumped a bear, probably a yearling from the way it ran; for at this season a yearling or a two-year-old will run almost like a deer, keeping to the thick cane as long as it can and then bolting across through the bushes of the ordinary swamp-land until it can reach another canebrake. After a three hours' run this particular animal managed to get clear away without one of the hunters ever seeing it, and it ran until all the dogs were tired out. A day or two afterward one of the other members of the party shot a small yearling; that is, a bear which would have been two years old in the following February. It was very lean, weighing about fifty-five pounds. The finely chewed acorns in its stomach showed that it was already beginning to find mast.

We had seen the tracks of an old she in the neighborhood, and the next morning we started to hunt her out. I went with Clive Metcalf. We had been joined overnight by Mr. Ichabod Osborn and his son Tom, two Louisiana planters, with six or eight hounds—or rather bear-dogs, for in these packs most of the animals are of mixed blood, and, as with all packs that are used in the genuine hunting of the wilderness, pedigree counts for nothing as compared with steadiness, courage, and intelligence. There were only two of the new dogs that were really stanch bear-dogs. The father of Ichabod Osborn had taken up the plantation upon which they were living in 1811, only a few years after Louisiana became part of the United States, and young Osborn was now the third in line from father to son who had steadily hunted bears in this immediate neighborhood.

On reaching the cypress slough near which the tracks of the old she had been seen the day before, Clive Metcalf and I separated from the others and rode off at a lively pace between two of the canebrakes. After an hour or two's wait we heard, very far off, the notes of one of the loudest-mouthed hounds, and instantly rode toward it, until we could make out the babel of the pack. Some hard galloping brought us opposite the point toward which they were heading—for experienced hunters can often tell the probable line of a bear's flight, and the spots at which it will break cover. But on this occasion the bear shied off from leaving the thick cane and doubled back; and soon the hounds were once more out of hearing, while we galloped desperately around the edge of the cane. The tough woods horses kept their feet like cats as they leaped logs, plunged through bushes, and dodged in and out among the tree-trunks; and we had all we could do to prevent the vines from lifting us out of the saddle, while the thorns tore our hands and faces. Hither and thither we went, now at a trot, now at a run, now stopping to listen for the pack. Occasionally we could hear the hounds, and then off we would go racing through the forest toward the point for which we thought they were heading. Finally, after a couple of hours of this, we came up on one side of a canebrake on the other side of which we could hear not only the pack but the yelling and cheering of Harley Metcalf and Tom Osborn and one or two of the negro hunters, all of whom were trying to keep the dogs up to their work in the thick cane. Again we rode ahead, and now in a few minutes were rewarded by hearing the leading dogs come to bay in the thickest of the cover. Having galloped as near to the spot as we could, we threw ourselves off the horses and plunged into the cane, trying to cause as little disturbance as possible, but of course utterly unable to avoid making some noise. Before we were within gunshot, however, we could tell by the sounds that the bear had once again started, making what is called a "walking bay." Clive Metcalf, a finished bear-hunter, was speedily able to determine what the bear's probable course would be, and we stole through the cane until we came to a spot near which he thought the quarry would pass. Then we crouched down, I with my rifle at the ready. Nor did we have long to wait. Peering through the thick-growing stalks I suddenly made out the dim outline of the bear coming straight toward us; and noiselessly I cocked and half raised my rifle, waiting for a clearer chance. In a few seconds it came; the bear turned almost broadside to me, and walked forward very stiff-legged, almost as if on tiptoe, now and then looking back at the nearest dogs. These were two in number—Rowdy, a very deep-voice hound, in the lead, and Queen, a shrill-tongued brindled bitch, a little behind. Once or twice the bear paused as she looked back at them, evidently hoping that they would come so near that by a sudden race she could catch one of them. But they were too wary.

All of this took but a few moments, and as I saw the bear quite distinctly some twenty yards off, I fired for behind the shoulder. Although I could see her outline, yet the cane was so thick that my sight was on it and not on the bear itself. But I knew my bullet would go true; and sure enough, at the crack of the rifle the bear stumbled and fell forward, the bullet having passed through both lungs and out at the opposite side. Immediately the dogs came running forward at full speed, and we raced forward likewise lest the pack should receive damage. The bear had but a minute or two to live, yet even in that time more than one valuable hound might lose its life; so when within half a dozen steps of the black, angered beast, I fired again, breaking the spine at the root of the neck; and down went the bear stark dead, slain in the canebrake in true hunter fashion. One by one the hounds struggled up and fell on their dead quarry, the noise of the worry filling the air. Then we dragged the bear out to the edge of the cane, and my companion wound his horn to summon the other hunters.

This was a big she bear, very lean, and weighing two hundred and two pounds. In her stomach were palmetto-berries, beetles, and a little mutton-cane, but chiefly acorns chewed up in a fine brown mass.

John McIlhenny had killed a she bear about the size of this on his plantation at Avery's Island the previous June. Several bear had been raiding his cornfields, and one evening he determined to try to waylay them. After dinner he left the ladies of his party on the gallery of his house while he rode down in a hollow and concealed himself on the lower side of the corn-field. Before he had waited ten minutes a she bear and her cub came into the field. The she rose on her hind legs, tearing down an armful of ears of corn which she seemingly gave to the cub, and then rose for another armful. McIlhenny shot her; tried in vain to catch the cub; and rejoined the party on the veranda, having been absent but one hour.

After the death of my bear I had only a couple of days left. We spent them a long distance from camp, having to cross two bayous before we got to the hunting-grounds. I missed a shot at a deer, seeing little more than the flicker of its white tail through the dense bushes; and the pack caught and killed a very lean two-year-old bear weighing eighty pounds. Near a beautiful pond called Panther Lake we found a deer-lick, the ground not merely bare, but furrowed into hollows by the tongues of the countless generations of deer that had frequented the place. We also passed a huge mound, the only hillock in the entire district; it was the work of man, for it had been built in the unknown past by those unknown people whom we call mound-builders. On the trip, all told, we killed and brought into camp three bear, six deer, a wildcat, a turkey, a possum, and a dozen squirrels; and we ate everything except the wildcat.

In the evenings we sat around the blazing camp-fires, and, as always on such occasions, each hunter told tales of his adventures and of the strange feats and habits of the beasts of the wilderness. There had been beaver all through this delta in the old days, and a very few are still left in out-of-the-way places. One Sunday morning we saw two wolves, I think young of the year, appear for a moment on the opposite side of the bayou, but they vanished before we could shoot. All of our party had had a good deal of experience with wolves. The Metcalfs had had many sheep killed by them, the method of killing being invariably by a single bite which tore open the throat while the wolf ran beside his victim. The wolves also killed young hogs, but were very cautious about meddling with an old sow; while one of the big half-wild boars that ranged free through the woods had no fear of any number of wolves. Their endurance and the extremely difficult nature of the country made it difficult to hunt them, and the hunters all bore them a grudge, because if a hound got lost in a region where wolves were at all plentiful they were almost sure to find and kill him before he got home. They were fond of preying on dogs, and at times would boldly kill the hounds right ahead of the hunters. In one instance, while the dogs were following a bear and were but a couple of hundred yards in front of the horsemen, a small party of wolves got in on them and killed two. One of the Osborns, having a valuable hound which was addicted to wandering in the woods, saved him from the wolves by putting a bell on him. The wolves evidently suspected a trap and would never go near the dog. On one occasion another of his hounds got loose with a chain on, and they found him a day or two afterward unharmed, his chain having become entangled in the branches of a bush. One or two wolves had evidently walked around and around the imprisoned dog, but the chain had awakened their suspicions and they had not pounced on him. They had killed a yearling heifer a short time before, on Osborn's plantation, biting her in the hams. It has been my experience that foxhounds as a rule are afraid of attacking a wolf; but all of my friends assured me that their dogs, if a sufficient number of them were together, would tackle a wolf without hesitation; the packs, however, were always composed, to the extent of at least half, of dogs which, though part hound, were part shepherd or bull of some other breed. Doctor Miller had hunted in Arkansas with a pack specially trained after the wolf. There were twenty-eight of them all told, and on this hunt they ran down and killed unassisted four full-grown wolves, although some of the hounds were badly cut. None of my companions had ever known of wolves actually molesting men, but Mr. Ichabod Osborn's son-in-law had a queer adventure with wolves while riding alone through the woods one late afternoon. His horse acting nervously, he looked about and saw that five wolves were coming toward him. One was a bitch, the other four were males. They seemed to pay little heed to him, and he shot one of the males, which crawled off. The next minute the bitch ran straight toward him and was almost at his stirrup when he killed her. The other three wolves, instead of running away, jumped to and fro, growling, with their hair bristling, and he killed two of them; whereupon the survivor at last made off. He brought the scalps of the three dead wolves home with him.

Near our first camp was the carcass of a deer, a yearling buck, which had been killed by a cougar. When first found, the wounds on the carcass showed that the deer had been killed by a bite in the neck at the back of the head; but there were scratches on the rump as if that panther had landed on its back. One of the negro hunters, Brutus Jackson, evidently a trustworthy man, told me that he had twice seen cougars, each time under unexpected conditions.

Once he saw a bobcat race up a tree, and riding toward it saw a panther reared up against the trunk. The panther looked around at him quite calmly, and then retired in leisurely fashion. Jackson went off to get some hounds and when he returned two hours afterward the bobcat was still up the tree, evidently so badly scared that he did not wish to come down. The hounds were unable to follow the cougar. On another occasion he heard a tremendous scuffle and immediately afterward saw a big doe racing along with a small cougar literally riding it. The cougar was biting the neck, but low down near the shoulders; he was hanging on with his front paws, but was tearing away with his hind claws, so that the deer's hair appeared to fill the air. As soon as Jackson appeared the panther left the deer. He shot it, and the doe galloped off, apparently without serious injury.

# 5

# Down an Unknown River into the Equatorial Forest

THE MIGHTIEST RIVER IN THE WORLD IS THE AMAZON. IT runs from west to east, from the sunset to the sunrise, from the Andes to the Atlantic. The main stream flows almost along the equator, while the basin which contains its affluents extends many degrees north and south of the equator. The gigantic equatorial river-basin is filled with an immense forest, the largest in the world, with which no other forest can be compared save those of western Africa and Malaysia. We were within the southern boundary of this great equatorial forest, on a river which was not merely unknown but unguessed at, no geographer having ever suspected its existence. This river flowed northward toward the equator, but whither it would go, whether it would turn one way or another, the length of its course, where it would come out, the character of the stream itself, and the character of the dwellers along its banks—all these things were yet to be discovered.

One morning while the canoes were being built Kermit and I walked a few kilometres down the river and surveyed the next rapids below. The vast still forest was almost empty of life. We found old Indian signs. There were very few birds and these in the tops of the tall trees. We saw a recent tapir track; and under a cajazeira-tree by the bank there were the tracks of capybaras which had been eating the fallen fruit. This fruit is delicious and would make a valuable

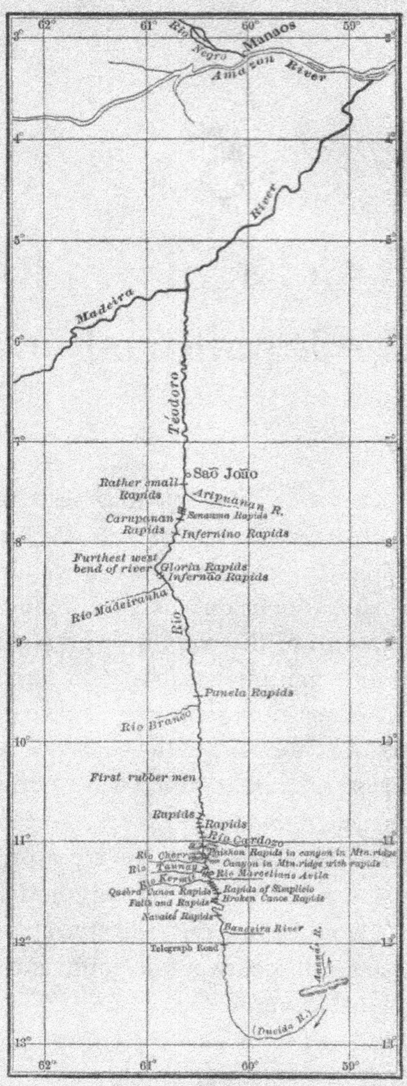


Sketch map of the unknown river christened Rio Roosevelt, and subsequently Rio Téodoro, by direction of the Brazilian Government

This map was prepared by Colonel Roosevelt from his journal and the diaries of Cherrie and of Kermit Roosevelt, the war having prevented the arrival of the map prepared by Lieutenant Lyra
The Ananás may be the headwaters of the Cardozo or of the Aripuanan, or it may flow into the Canuma or Tapajos; it will not be put on the map until it is actually descended

**Fig. 1.** Sketch map and explanatory caption as they appeared in Theodore Roosevelt, *Through the Brazilian Wilderness* (Charles Scribner's Sons, 1914).

addition to our orchards. The tree although tropical is hardy, thrives when domesticated, and propagates rapidly from shoots.

The Department of Agriculture should try whether it would not grow in southern California and Florida. This was the tree from which the doctor's family name was taken. His parental grandfather, although of Portuguese blood, was an intensely patriotic Brazilian. He was a very young man when the independence of Brazil was declared, and did not wish to keep the Portuguese family name; so he changed it to that of the fine Brazilian tree in question. Such change of family name is common in Brazil. Doctor Vital Brazil, the student of poisonous serpents, was given his name by his father, whose own family name was entirely different; and his brother's name was again different.

There were tremendous downpours of rain, lasting for a couple of hours and accompanied by thunder and lightning. But on the whole it seemed as if the rains were less heavy and continuous than they had been. We all of us had to help in building the canoes now and then. Kermit, accompanied by Antonio the Parecís and João, crossed the river and walked back to the little river that had entered from the east, so as to bring back a report of it to Colonel Rondon. Lyra took observations, by the sun and by the stars. We were in about latitude eleven degrees twenty-one minutes south, and due north of where we had started. The river had wound so that we had gone two miles for every one we made northward. Our progress had been very slow; and until we got out of the region of incessant rapids, with their attendant labor and hazard, it was not likely that we should go much faster.

On the morning of March 22 we started in our six canoes. We made ten kilometres. Twenty minutes after starting we came to the first rapids. Here every one walked except the three best paddlers, who took the canoes down in succession—an hour's job. Soon after this we struck a bees' nest in the top of a tree overhanging the river; our steersman climbed out and robbed it, but, alas! Lost the honey on the way back. We came to a small steep fall which we did not dare run in our overladen, clumsy, and cranky dugouts. Fortunately, we were able to follow a deep canal which led off for a kilometre, returning just below the falls, fifty yards from where it had started. Then, having been in the boats and in motion only one hour and a half, we came to a long stretch of rapids which it took us six hours to descend, and we camped at the foot. Everything was taken out of the canoes, and they were run down in succession. At one difficult and perilous place they were let down by ropes; and even thus we almost lost one.

We went down the right bank. On the opposite bank was an Indian village, evidently inhabited only during the dry season. The marks on the stumps of trees showed that these Indians had axes and knives; and there were old fields in which maize, beans, and cotton had been grown. The forest dripped and steamed. Rubber-trees were plentiful. At one point the tops of a group of tall trees were covered with yellow-white blossoms. Others bore red blossoms. Many of the big trees, of different kinds, were buttressed at the base with great thin walls of wood. Others, including both palms and ordinary trees, showed an even stranger peculiarity. The trunk, near the base, but sometimes six or eight feet from the ground, was split into a dozen or twenty branches or small trunks which sloped outward in tent-like shape, each becoming a root. The larger trees of this type looked as if their trunks were seated on the tops of the pole frames of Indian tepees. At one point in the stream, to our great surprise, we saw a flying-fish. It skimmed the water like a swallow for over twenty yards.

Although we made only ten kilometres we worked hard all day. The last canoes were brought down and moored to the bank at nightfall. Our tents were pitched in the darkness.

Next day we made thirteen kilometres. We ran, all told, a little over an hour and three-quarters. Seven hours were spent in getting past a series of rapids at which the portage, over rocky and difficult ground, was a kilometre long. The canoes were run down empty—a hazardous run, in which one of them upset.

Yet while we were actually on the river, paddling and floating downstream along the reaches of swift, smooth water, it was very lovely. When we started in the morning the day was overcast and the air was heavy with vapor. Ahead of us the shrouded river stretched between dim walls of forest, half seen in the mist. Then the sun burned up the fog and loomed through it in a red splendor that changed first to gold and then to molten white. In the dazzling light, under the brilliant blue of the sky, every detail of the magnificent forest was vivid to the eye: the great trees, the network of bush-ropes, the caverns of greenery where thick-leaved vines covered all things else. Wherever there was a hidden boulder the surface of the current was broken by waves. In one place, in midstream, a pyramidal rock thrust itself six feet above the surface of the river. On the banks we found fresh Indian sign.

At home in Vermont, Cherrie is a farmer, with a farm of six hundred acres, most of it woodland. As we sat at the foot of the rapids, watching for the last dugouts with their naked paddlers to swing into sight round the bend through the white water, we talked of the northern spring that was just beginning, He sells cream, eggs, poultry, potatoes, honey, occasionally pork and veal; but at this season it was the time for the maple-sugar crop. He has a sugar-orchard, where he taps twelve hundred trees and hopes soon to tap as many more in addition. Said Cherrie: "It's a busy time now for Fred Rice"—Fred Rice is the hired man, and in sugar-time the Cherrie boys help him with enthusiasm, and, moreover, are paid with exact justice for the work they do. There is much wild life about the farm, although it is near Brattleboro. One night in early spring a bear left his tracks near the sugar-house; and now and then in summer Cherrie has had to sleep in the garden to keep the deer away from the beans, cabbages, and beets.

There was not much bird life in the forest, but Cherrie kept getting species new to the collection. At this camp he shot an interesting little ant-thrush. It was the size of a warbler, jet-black, with white under-surfaces of the wings and tail, white on the tail-feathers, and a large spot of white on the back, normally almost concealed, the feathers on the back being long and fluffy. When he shot the bird, a male, it was showing off before a dull-colored little bird, doubtless the female; and the chief feature of the display was this white spot on the back. The white feathers were raised and displayed so that the spot flashed like the "chrysanthemum" on a prongbuck whose curiosity has been aroused. In the gloom of the forest the bird was hard to see, but the flashing of this patch of white feathers revealed it at once, attracting immediate attention. It was an excellent example of a coloration mark which served a purely advertising purpose; apparently it was part of a courtship display, the bird was about thirty feet up in the branches.

In the morning, just before leaving this camp, a tapir swam across stream a little way above us; but unfortunately we could not get a shot at it. An ample supply of tapir beef would have meant much to us. We had started with fifty days' rations; but this by no means meant full rations, in the sense of giving every man all he wanted to eat. We had two meals a day, and were on rather short commons—both our mess and the camaradas'—except when we got plenty of palm tops. For our mess we had the boxes chosen by Fiala, each containing a day's rations for six men, our number. But we made each box last a day and a half, or at times two days, and in addition we gave some of the food to the camaradas. It was only on the rare occasions when we had killed some monkeys or curassows, or caught some fish, that everybody had enough. We would have welcomed that tapir. So far the game, fish, and fruit had been too scarce to be an element of weight in our food-supply. In an exploring trip like ours, through a difficult and utterly unknown country, especially if densely forested, there is little time to halt, and game cannot be counted on. It is only in lands like our own West thirty years ago, like South Africa in the middle of the last century, like East Africa today that game can be made the chief food-supply. On this trip our only substantial food-supply from the country hitherto had been that furnished by the palm tops. Two men were detailed every day to cut down palms for food.

A kilometre and a half after leaving this camp we came on a stretch of big rapids. The river here twists in loops, and we had heard the roaring of these rapids the previous afternoon. Then we passed out of earshot of them; but Antonio Correa, our best waterman, insisted all along that the roaring meant rapids worse than any we had encountered for some days. "I was brought up in the water, and I know it like a fish, and all its sounds," said he. He was right. We had to carry the loads nearly a kilometre that afternoon, and the canoes were pulled out on the bank so that they might be in readiness to be dragged overland next day. Rondon, Lyra, Kermit, and Antonio Correa explore both sides of the river. On the opposite or left bank they found the mouth of a considerable river, bigger than the Rio Kermit, flowing in from the west and making its entrance in the middle of the rapids. This river we christened the Taunay, in honor of a distinguished Brazilian, an explorer, a soldier, a senator, who was also a writer of note. Kermit had with him two of his novels, and I had read one of his books dealing with a disastrous retreat during the Paraguayan war.

Next morning, the 25th, the canoes were brought down. A path was chopped for them and rollers laid; and half-way down the rapids Lyra and Kermit, who were overseeing the work as well as doing their share of the pushing and hauling, got them into a canal of smooth water, which saved much severe labor. As our food-supply lowered we were constantly more desirous of economizing the strength of the men. One day more would complete a month since we had embarked on the Dúvida—as we had started in February, the lunar and calendar months coincided. We had used up over half our provisions. We had come only a trifle over one hundred and sixty kilometres, thanks to the character and number of the rapids. We believed we had three or four times the distance yet to go before coming to a part of the river where we might hope to meet assistance, either from rubber-gatherers or from Pyrineus, if he were really coming up the river which we were going down. If the rapids continued to be as they had been it could not be much more than three weeks before we were in straits for food, aside from the ever-present danger of accident in the rapids; and if our progress were no faster than it had been—and we were straining to do our best—we would in such event still have several hundreds of kilometres of unknown river before us. We could not even hazard a guess at what was in front. The river was now a really big river, and it seemed impossible that it could flow either into the Gy-Paraná or the Tapajos. It was possible that it went into the Canumá, a big affluent of the Madeira low down and next to the Tapajos. It was more probable that it was the headwaters of the Aripuanan, a river which, as I have said, was not even named on the excellent English map of Brazil I carried. Nothing but the mouth had been known to any geographer; but the lower course had long been known to rubber-gatherers, and recently a commission from the government of Amazonas had part way ascended one branch of it—not as far as the rubber-gatherers had gone, and, as it turned out, not the branch we came down.

Two of our men were down with fever. Another man, Julio, a fellow of powerful frame, was utterly worthless, being an inborn, lazy shirk with the heart of a ferocious cur in the body of a bullock.

The others were good men, some of them very good indeed. They were under the immediate supervision of Pedrinho Craveiro, who was first-class in every way.

This camp was very lovely. It was on the edge of a bay, into which the river broadened immediately below the rapids. There was a beach of white sand, where we bathed and washed our clothes. All around us, and across the bay, and on both sides of the long water-street made by the river, rose the splendid forest. There were flocks of parakeets colored green, blue, and red. Big toucans called overhead, lustrous green-black in color, with white throats, red gorgets, red-and-yellow tail-coverts, and huge black-and-yellow bills. Here the soil was fertile; it will be a fine site for a coffee-plantation when this region is open to settlement. Surely such a rich and fertile land cannot be permitted to remain idle, to lie as a tenantless wilderness, while there are such teeming swarms of human beings in the overcrowded, over-peopled countries of the Old World. The very rapids and waterfalls which now make the navigation of the river so difficult and dangerous would drive electric trolleys up and down its whole length and far out on either side, and run mills and factories, and lighten the labor on farms. With the incoming of settlement and with the steady growth of knowledge how to fight and control tropical diseases, fear of danger to health would vanish. A land like this is a hard land for the first explorers, and perhaps for their immediate followers, but not for the people who come after them.

In mid-afternoon we were once more in the canoes; but we had paddled with the current only a few minutes, we had gone only a kilometre, when the roar of rapids in front again forced us to haul up to the bank. As usual, Rondon, Lyra, and Kermit, with Antonio Correa, explored both sides while camp was being pitched. The rapids were longer and of steeper descent than the last, but on the opposite or western side there was a passage down which we thought we could get the empty dugouts at the cost of dragging them only a few yards at one spot. The loads were to be carried down the hither bank, for a kilometre, to the smooth water. The river foamed between great rounded masses of rock, and at one point there was a sheer fall of six or eight feet. We found and ate wild pineapples. Wild beans were in flower. At dinner we had a toucan and a couple of parrots, which were very good.

All next day was spent by Lyra in superintending our three best watermen as they took the canoes down the west side of the rapids, to the foot, at the spot to which the camp had meantime been shifted. In the forest, some of the huge sipas, or rope vines, which were as big as cables, bore clusters of fragrant flowers. The men found several honey-trees, and fruits of various kinds, and small cocoanuts; they chopped down an ample number of palms, for the palm-cabbage; and, most important of all, they gathered a quantity of big Brazil-nuts, which when roasted tasted like the best of chestnuts and are nutritious; and they caught a number of big piranhas, which were good eating. So we all had a feast, and everybody had enough to eat and was happy.

By these rapids, at the fall, Cherrie found some strange carvings on a bare mass of rock. They were evidently made by men a long time ago. As far as is known, the Indians thereabouts make no such figures now. They were in two groups, one on the surface of the rock facing the land, the other on that facing the water. The latter were nearly obliterated. The former were in good preservation, the figures sharply cut into the rock. They consisted, upon the upper flat part of the rock, of four multiple circles with a dot in the middle (◎) very accurately made and about a foot and a half in diameter; and below them, on the side of the rock, four multiple m's or inverted w's (⩕). What these curious symbols represented or who made them, we could not, of course, form the slightest idea. It may be that in a very remote past some Indian tribes of comparatively advanced culture had penetrated to this lovely river, just as we had now come to it. Before white men came to South America there had already existed therein various semicivilizations, some rude, others fairly advanced, which rose, flourished, and persisted through immemorial ages, and then vanished. The vicissitudes in the history of humanity during its stay on this southern continent have been as strange, varied, and inexplicable as paleontology shows to have been the case, on the same continent, in the history of the higher forms of animal life during the age of mammals. Colonel Rondon stated that such figures as these are not found anywhere else in Matto Grosso where he has been, and therefore it was all the more strange to find them in this one place on the unknown river, never before visited by white men, which we were descending.

Next morning we went about three kilometres before coming to some steep hills, beautiful to look upon, clad as they were in dense, tall, tropical forest, but ominous of new rapids. Sure enough, at their foot we had to haul up and prepare for a long portage. The canoes we ran down empty. Even so, we were within an ace of losing two, the lashed couple in which I ordinarily journeyed. In a sharp bend of the rapids, between two big curls, they were swept among the boulders and under the matted branches which stretched out from the bank. They filled, and the racing current pinned them where they were, one partly on the other. All of us had to help get them clear. Their fastenings were chopped asunder with axes. Kermit and half a dozen of the men, stripped to the skin, made their way to a small rock island in the little falls just above the canoes, and let down a rope which we tied to the outer-most canoe. The rest of us, up to our armpits and barely able to keep our footing as we slipped and stumbled among the boulders in the swift current, lifted and shoved while Kermit and his men pulled the rope and fastened the slack to a half-submerged tree. Each canoe in succession was hauled up the little rock island, baled, and then taken down in safety by two paddlers. It was nearly four o'clock before we were again ready to start, having been delayed by a rain-storm so heavy that we could not see across the river. Ten minutes' run took us to the head of another series of rapids; the exploring party returned with the news that we had an all-day's job ahead of us; and we made camp in the rain, which did not matter much, as we were already drenched through. It was impossible, with the wet wood, to make a fire sufficiently hot to dry all our soggy things, for the rain was still falling. A tapir was seen from our boat, but, as at the moment we were being whisked round in a complete circle by a whirlpool, I did not myself see it in time to shoot.

Next morning we went down a kilometre, and then landed on the other side of the river. The canoes were run down, and the loads carried to the other side of a little river coming in from the west, which Colonel Rondon christened Cherrie River. Across this we went on a bridge consisting of a huge tree felled by Macairo, one of our best men. Here we camped while Rondon, Lyra, Kermit, and Antonio Correa explored what was ahead. They were absent until mid-afternoon. Then they returned with the news that we were among ranges of low mountains, utterly different in formation from the high plateau region to which the first rapids, those we had come to on the 2d of March, belonged. Through the first range of these mountains the river ran in a gorge, some three kilometres long, immediately ahead of us. The ground was so rough and steep that it would be impossible to drag the canoes over it and difficult enough to carry the loads; and the rapids were so bad, containing several falls, one of at least ten metres in height, that it was doubtful how many of the canoes we could get down them. Kermit, who was the only man with much experience of rope work, was the only man who believed we could get the canoes down at all; and it was, of course, possible that we should have to build new ones at the foot to supply the place of any that were lost or left behind. In view of the length and character of the portage and of all the unpleasant possibilities that were ahead and of the need of keeping every pound of food, it was necessary to reduce weight in every possible way and to throw away everything except the barest necessities.

We thought we had reduced our baggage before; but now we cut to the bone. We kept the fly for all six of us to sleep under. Kermit's shoes had gone, thanks to the amount of work in the water which he had been doing; and he took the pair I had been wearing, while I put on my spare pair. In addition to the clothes I wore, I kept one set of pajamas, a spare pair of drawers, a spare pair of socks, half a dozen handkerchiefs, my wash-kit, my pocket medicine case, and a little bag containing my spare spectacles, gun grease, some adhesive plaster, some needles and thread, the "fly dope," and my purse and letter of credit, to be used at Manaos. All of these went into the bag containing my cot, blanket, and mosquito-net. I also carried a cartridge bag containing my cartridges, head net, and gantlets. Kermit cut down even closer; and the others about as close.

The last three days of March we spent in getting to the foot of the rapids in this gorge. Lyra and Kermit, with four of the best watermen, handled the empty canoes. The work was not only difficult and laborious in the extreme, but hazardous; for the walls of the gorge were so sheer that at the worst places they had to cling to narrow shelves on the face of the rock, while letting the canoes down with ropes. Meanwhile Rondon surveyed and cut a trail for the burden-bearers, and superintended the portage of the loads. The rocky sides of the gorge were too steep for laden men to attempt to traverse them. Accordingly the trail had to go over the top of the mountain, both the ascent and the descent of the rock-strewn, forest-clad slopes being very steep. It was hard work to carry loads over such a trail. From the top of the mountain, through an opening in the trees on the edge of a cliff, there was a beautiful view of the country ahead. All around and in front of us there were ranges of low mountains about the height of the lower ridges of the Alleghanies. Their sides were steep and they were covered with the matted growth of the tropical forest. Our next camping-place, at the foot of the gorge, was almost beneath us, and from thence the river ran in a straight line, flecked with white water, for about a kilometre. Then it disappeared behind and between mountain ridges, which we supposed meant further rapids. It was a view well worth seeing; but, beautiful although the country ahead of us was, its character was such as to promise further hardships, difficulty, and exhausting labor, and especially further delay; and delay was a serious matter to men whose food-supply was beginning to run short, whose equipment was reduced to the minimum, who for a month, with the utmost toil, had made very slow progress, and who had no idea of either the distance or the difficulties of the route in front of them.

There was not much life in the woods, big or little. Small birds were rare, although Cherrie's unwearied efforts were rewarded from time to time by a species new to the collection. There were tracks of tapir, deer, and agouti; and if we had taken two or three days to devote to nothing else than hunting them we might perchance have killed something; but the chance was much too uncertain, the work we were doing was too hard and wearing, and the need of pressing forward altogether too great to permit us to spend any time in such manner. The hunting had to come in incidentally. This type of well-nigh impenetrable forest is the one in which it is most difficult to get even what little game exists therein. A couple of curassows and a big monkey were killed by the colonel and Kermit. On the day the monkey was brought in Lyra, Kermit and their four associates had spent from sunrise to sunset in severe and at moments dangerous toil among the rocks and in the swift water, and the fresh meat was appreciated. The head, feet, tail, skin, and entrails were boiled for the gaunt and ravenous dogs. The flesh gave each of us a few mouthfuls; and how good those mouthfuls tasted!

Cherrie, in addition to being out after birds in every spare moment, helped in all emergencies. He was a veteran in the work of the tropic wilderness. We talked together often, and of many things, for our views of life, and of a man's duty to his wife and children, to other men and to women, and to the State in peace and war, were in all essentials the same. His father had served all through the Civil War, entering an Iowa cavalry regiment as a private and coming out as a captain; his breast-bone was shattered by a blow from a musket butt, in hand-to-hand fighting at Shiloh.

During this portage the weather favored us. We were coming toward the close of the rainy season. On the last day of the month, when we moved camp to the foot of the gorge, there was a thunder-storm; but on the whole we were not bothered by rain until the last night, when it rained heavily, driving under the fly so as to wet my cot and bedding. However, I slept comfortably enough, rolled in the damp blanket. Without the blanket I should have been uncomfortable; a blanket is a necessity for health. On the third day Lyra and Kermit, with their daring and hard-working watermen, after wearing labor, succeeded in getting five canoes through the worst of the rapids to the chief fall. The sixth, which was frail and weak, had its bottom beaten out on the jagged rocks of the broken water. On this night, although I thought I had put my clothes out of reach, both the termites and the carregadores ants got at them, ate holes in one boot, ate one leg of my drawers, and riddled my handkerchief; and I now had nothing to replace anything that was destroyed.

Next day Lyra, Kermit, and their camaradas brought the five canoes that were left down to camp. They had in four days accomplished a work of incredible labor and of the utmost importance; for at the first glance it had seemed an absolute impossibility to avoid abandoning the canoes when we found that the river sank into a cataract-broken torrent at the bottom of a canyon-like gorge between steep mountains. On April 2 we once more started, wondering how soon we should strike other rapids in the mountains ahead, and whether in any reasonable time we should, as the aneroid indicated, be so low down that we should necessarily be in a plain where we could make a journey of at least a few days without rapids. We had been exactly a month going through an uninterrupted succession of rapids. During that month we had come only about one hundred and ten kilometres, and had descended nearly one hundred and fifty metres—the figures are approximate but fairly accurate.[1] We had lost four of the canoes with which we started, and one other, which we had built, and the life of one man; and the life of a dog which by its death had in all probability saved the life of Colonel Rondon. In a straight line northward toward our supposed destination, we had not made more than a mile and a quarter a day; at the cost of bitter toil for most of the party, of much risk for some of the party, and of some risk and some hardship for all of the party. Most of the camaradas were downhearted, naturally enough, and occasionally asked one of us if we really believed that we should ever get out alive; and we had to cheer them up as best we could.

There was no change in our work for the time being. We made but three kilometres that day. Most of the party walked all the time; but the dugouts carried the luggage until we struck the head of the series of rapids which were to take up the next two or three days. The river rushed through a wild gorge, a chasm or canyon, between two mountains. Its sides were very steep, mere rock walls, although in most places so covered with the luxuriant growth of the trees and bushes that clung in the crevices and with green moss that the naked rock was hardly seen. Rondon, Lyra, and Kermit, who were in front, found a small level spot with a beach of sand, and sent back word to camp there while they spent several hours in exploring the country ahead. The canoes were run down empty, and the loads carried painfully along the face of the cliffs; so bad was the trail that I found it rather hard to follow, although carrying nothing but my rifle and cartridge bag. The explorers returned with the information that the mountains stretched ahead of us, and that there were rapids as far as they had gone. We could only hope that the aneroid was not hopelessly out of kilter, and that we should, therefore, fairly soon find ourselves in comparatively level country. The severe toil, on a rather limited food-supply, was telling on the strength as well as on the spirits of the men; Lyra and Kermit, in addition to their other work, performed as much actual physical labor as any of them.

Next day, the 3d of April, we began the descent of these sinister rapids of the chasm. Colonel Rondon had gone to the summit of the mountain in order to find a better trail for the burden-bearers, but it was hopeless, and they had to go along the face of the cliffs. Such an exploring expedition as that in which we were engaged of necessity involves hard and dangerous labor and perils of many kinds. To follow downstream an unknown river, broken by innumerable cataracts and rapids, rushing through mountains of which the existence has never even been guessed, bears no resemblance whatever to following even a fairly dangerous river which has been thoroughly explored and has become in some sort a highway, so that experienced pilots can be secured as guides, while the portages have been pioneered and trails chopped out, and every dangerous feature of the rapids is known beforehand. In this case no one could foretell that the river would cleave its way through steep mountain chains, cutting narrow clefts in which the cliff walls rose almost sheer on either hand. When a rushing river thus "canyons," as we used to say out West, and the mountains are very steep, it becomes almost impossible to bring the canoes down the river itself and utterly impossible to portage them along the cliff sides, while even to bring the loads over the mountain is a task of extraordinary labor and difficulty. Moreover, no one can tell how many times the task will have to be repeated or when it will end or whether the food will hold out; every hour of work in the rapids is fraught with the possibility of the gravest disaster, and yet it is imperatively necessary to attempt it; and all this is done in an uninhabited wilderness, or else a wilderness tenanted only by unfriendly savages, where failure to get through means death by disease and starvation. Wholesale disasters to South American exploring parties have been frequent. The first recent effort to descend one of the unknown rivers to the Amazon from the Brazilian highlands resulted in such a disaster. It was undertaken in 1889 by a party about as large as ours under a Brazilian engineer officer, Colonel Telles Peres. In descending some rapids they lost everything—canoes, food, medicine, implements—everything. Fever smote them, and then starvation. All of them died except one officer and two men who were rescued months later. Recently, in Guiana, a wilderness veteran, André, lost two-thirds of his party by starvation. Genuine wilderness exploration is as dangerous as warfare. The conquest of wild nature demands the utmost vigor, hardihood, and daring, and takes from the conquerors a heavy toll of life and health.

Lyra, Kermit, and Cherrie, with four of the men, worked the canoes half-way down the canyon. Again and again it was touch and go whether they could get by a given point. At one spot the channel of the furious torrent was only fifteen yards across. One canoe was lost, so that of the seven with which we had started only two were left. Cherrie labored with the other men at times, and also stood as guard over them, for, while actually working, of course no one could carry a rifle. Kermit's experience in bridge-building was invaluable in enabling him to do the rope work by which alone it was possible to get the canoes down the canyon. He and Lyra had now been in the water for days. Their clothes were never dry. Their shoes were rotten. The bruises on their feet and legs had become sores. On their bodies some of the insect bites had become festering wounds, as indeed was the case with all of us. Poisonous ants, biting flies, ticks, wasps, bees were a perpetual torment. However, no one had yet been bitten by a venomous serpent, a scorpion, or a centipede, although we had killed all of the three within the camp limits.

Under such conditions whatever is evil in men's natures comes to the front. On this day a strange and terrible tragedy occurred. One of the camaradas, a man of pure European blood, was the man named Julio, of whom I have already spoken. He was a very powerful fellow and had been importunately eager to come on the expedition; and he had the reputation of being a good worker. But, like so many men of higher standing, he had had no idea of what such an expedition really meant, and under the strain of toil, hardship, and danger his nature showed its true depths of selfishness, cowardice, and ferocity. He shirked all work. He shammed sickness. Nothing could make him do his share; and yet unlike his self-respecting fellows he was always shamelessly begging for favors. Kermit was the only one of our party who smoked; and he was continually giving a little tobacco to some of the camaradas, who worked especially well under him. The good men did not ask for it; but Julio, who shirked every labor, was always, and always in vain, demanding it. Colonel Rondon, Lyra, and Kermit each tried to get work out of him, and in order to do anything with him had to threaten to leave him in the wilderness. He threw all his tasks on his comrades, and, moreover, he stole their food as well as ours. On such an expedition the theft of food comes next to murder as a crime, and should by rights be punished as such. We could not trust him to cut down palms or gather nuts, because he would stay out and eat what ought to have gone to the common store. Finally, the men on several occasions themselves detected him stealing their food. Alone of the whole party, and thanks to the stolen food, he had kept in full flesh and bodily vigor.

One of our best men was a huge negro named Paixão—Paishon—a corporal and acting sergeant in the engineer corps. He had, by the way, literally torn his trousers to pieces, so that he wore only the tatters of a pair of old drawers until I gave him my spare trousers when we lightened loads. He was a stern disciplinarian. One evening he detected Julio stealing food and smashed him in the mouth. Julio came crying to us, his face working with fear and malignant hatred; but after investigation he was told that he had gotten off uncommonly lightly. The men had three or four carbines, which were sometimes carried by those who were not their owners.

On this morning, at the outset of the portage, Pedrinho discovered Julio stealing some of the men's dried meat. Shortly afterward Paishon rebuked him for, as usual, lagging behind. By this time we had reached the place where the canoes were tied to the bank and then taken down one at a time. We were sitting down, waiting for the last loads to be brought along the trail. Pedrinho was still in the camp we had left. Paishon had just brought in a load, left it on the ground with his carbine beside it, and returned on the trail for another load. Julio came in, put down his load, picked up the carbine, and walked back on the trail, muttering to himself but showing no excitement. We thought nothing of it, for he was always muttering; and occasionally one of the men saw a monkey or big bird and tried to shoot it, so it was never surprising to see a man with a carbine.

In a minute we heard a shot; and in a short time three or four of the men came up the trail to tell us that Paishon was dead, having been shot by Julio, who had fled into the woods. Colonel Rondon and Lyra were ahead; I sent a messenger for them, directed Cherrie and Kermit to stay where they were and guard the canoes and provisions, and started down the trail with the doctor—an absolutely cool and plucky man, with a revolver but no rifle—and a couple of the camaradas. We soon passed the dead body of poor Paishon. He lay in a huddle, in a pool of his own blood, where he had fallen, shot through the heart. I feared that Julio had run amuck, and intended merely to take more lives before he died, and that he would begin with Pedrinho, who was alone and unarmed in the camp we had left. Accordingly I pushed on, followed by my companions, looking sharply right and left; but when we came to the camp the doctor quietly walked by me, remarking: "My eyes are better than yours, colonel; if he is in sight I'll point him out to you, as you have the rifle." However, he was not there, and the others soon joined us with the welcome news that they had found the carbine.

The murderer had stood to one side of the path and killed his victim, when a dozen paces off, with deliberate and malignant purpose. Then evidently his murderous hatred had at once given way to his innate cowardice, and, perhaps hearing some one coming along the path, he fled in panic terror into the wilderness. A tree had knocked the carbine from his hand. His footsteps showed that after going some rods he had started to return, doubtless for the carbine, but had fled again, probably because the body had then been discovered. It was questionable whether or not he would live to reach the Indian villages, which were probably his goal. He was not a man to feel remorse—never a common feeling; but surely that murderer was in a living hell, as, with fever and famine leering at him from the shadows, he made his way through the empty desolation of the wilderness. França, the cook, quoted out of the melancholy proverbial philosophy of the people of the proverb,

"No man knows the heart of any one"; and then expressed with deep conviction a weird ghostly belief I had never encountered before: "Paishon is following Julio now, and will follow him until he dies; Paishon fell forward on his hands and knees, and when a murdered man falls like that his ghost will follow the slayer as long as the slayer lives."

We did not attempt to pursue the murderer. We could not legally put him to death, although he was a soldier who in cold blood had just deliberately killed a fellow soldier. If we had been near civilization we would have done our best to bring him in and turn him over to justice. But we were in the wilderness, and how many weeks' journey was ahead of us we could not tell. Our food was running low, sickness was beginning to appear among the men, and both their courage and their strength were gradually ebbing. Our first duty was to save the lives and the health of the men of the expedition who had honestly been performing, and had still to perform, so much perilous labor. If we brought the murderer in he would have to be guarded night and day on an expedition where there were always loaded firearms about, and where there would continually be opportunity and temptation for him to make an effort to seize food and a weapon and escape, perhaps murdering some other good man. He could not be shackled while climbing along the cliff slopes; he could not be shackled in the canoes, where there was always chance of upset and drowning; and standing guard would be an additional and severe penalty on the weary, honest men already exhausted by overwork. The expedition was in peril, and it was wise to take every chance possible that would help secure success. Whether the murderer lived or died in the wilderness was of no moment compared with the duty of doing everything to secure the safety of the rest of the party. For the two days following we were always on the watch against his return, for he could have readily killed some one else by rolling rocks down on any of the men working on the cliff sides or in the bottom of the gorge. But we did not see him until the morning of the third day. We had passed the last of the rapids of the chasm, and the four boats were going downstream when he appeared behind some trees on the bank and called out that he wished to surrender and be taken aboard; for the murderer was an arrant craven at heart, a strange mixture of ferocity and cowardice. Colonel Rondon's boat was far in advance; he did not stop nor answer. I kept on in similar fashion with the rear boats, for I had no intention of taking the murderer aboard, to the jeopardy of the other members of the party, unless Colonel Rondon told me that it would have to be done in pursuance of his duty, as an officer of the army and a servant of the government of Brazil. At the first halt Colonel Rondon came up to me and told me that this was his view of his duty, but that he had not stopped because he wished first to consult me as the chief of the expedition. I answered that for the reasons enumerated above I did not believe that in justice to the good men of the expedition we should jeopardize their safety by taking the murderer along, and that if the responsibility were mine I should refuse to take him; but that he, Colonel Rondon, was the superior officer of both the murderer and of all the other enlisted men and army officers on the expedition, and in return was responsible for his actions to his own governmental superiors and to the laws of Brazil; and that in view of this responsibility he must act as his sense of duty bade him. Accordingly, at the next camp he sent back two men, expert woodsmen, to find the murderer and bring him in. They failed to find him.[2]

I have anticipated my narrative because I do not wish to recur the horror more than is necessary. I now return to my story. After we found that Julio had fled, we returned to the scene of the tragedy. The murdered man lay with a handkerchief thrown over his face. We buried him beside the place where he fell. With axes and knives the camaradas dug a shallow grave while we stood by with bared heads. Then reverently and carefully we lifted the poor body which but half an hour before had been so full of vigorous life. Colonel Rondon and I bore the head and shoulders. We laid him in the grave, and heaped a mound over him, and put a rude cross at his head. We fired a volley for a brave and loyal soldier who had died doing his duty. Then we left him forever, under the great trees beside the lonely river.

That day we got only half-way down the rapids. There was no good place to camp. But at the foot of one steep cliff there was a narrow, boulder-covered slope where it was possible to sling hammocks and cook; and a slanting spot was found for my cot, which had sagged until by this time it looked like a broken-backed centipede. It rained a little during the night, but not enough to wet us much. Next day Lyra, Kermit, and Cherrie finished their job, and brought the four remaining canoes to camp, one leaking badly from the battering on the rocks. We then went downstream a few hundred yards, and camped on the opposite side; it was not a good camping-place, but it was better than the one we left.

The men were growing constantly weaker under the endless strain of exhausting labor. Kermit was having an attack of fever, and Lyra and Cherrie had touches of dysentery, but all three continued to work. While in the water trying to help with an upset canoe I had by my own clumsiness bruised my leg against a boulder; and the resulting inflammation was somewhat bothersome. I now had a sharp attack of fever, but thanks to the excellent care of the doctor, was over it in about forty-eight hours; but Kermit's fever grew worse and he too was unable to work for a day or two. We could walk over the portages, however. A good doctor is an absolute necessity on an exploring expedition in such a country as that we were in, under penalty of a frightful mortality among the members; and the necessary risks and hazards are so great, the chances of disaster so large, that there is no warrant for increasing them by the failure to take all feasible precautions.

The next day we made another long portage round some rapids, and camped at night still in the hot, wet, sunless atmosphere of the gorge. The following day, April 6, we portaged past another set of rapids, which proved to be the last of the rapids of the chasm. For some kilometres we kept passing hills, and feared lest at any moment we might again find ourselves fronting another mountain gorge; with, in such case, further days of grinding and perilous labor ahead of us, while our men were disheartened, weak, and sick. Most of them had already begun to have fever. Their condition was inevitable after over a month's uninterrupted work of the hardest kind in getting through the long series of rapids we had just passed; and a long further delay, accompanied by wearing labor, would have almost certainly meant that the weakest among our party would have begun to die. There were already two of the camaradas who were too weak to help the others, their condition being such as to cause us serious concern.

However, the hills gradually sank into a level plain, and the river carried us through it at a rate that enabled us during the remainder of the day to reel off thirty-six kilometres, a record that for the first time held out promise. Twice tapirs swam the river while we passed, but not near my canoe. However, the previous evening Cherrie had killed two monkeys and Kermit one, and we all had a few mouthfuls of fresh meat; we had already had a good soup made out of a turtle Kermit had caught. We had to portage by one short set of rapids, the unloaded canoes being brought down without difficulty. At last, at four in the afternoon, we came to the mouth of a big river running in from the right. We thought it was probably the Ananás, but, of course, could not be certain. It was less in volume than the one we had descended, but nearly as broad; its breadth at this point being ninety-five yards as against one hundred and twenty for the larger river. There were rapids ahead, immediately after the junction, which took place in latitude ten degrees fifty-eight minutes south. We had come two hundred and sixteen kilometres all told, and were nearly north of where we had started. We camped on the point of land between the two rivers. It was extraordinary to realize that here about the eleventh degree we were on such a big river, utterly unknown to the cartographers and not indicated by even a hint on any map. We named this big tributary Rio Cardoza, after a gallant officer of the commission who had died of beriberi just as our expedition began. We spent a day at this spot, determining our exact position by the sun, and afterward by the stars, and sending on two men to explore the rapids in advance. They returned with the news that there were big cataracts in them, and that they would form an obstacle to our progress. They had also caught a huge siluroid fish, which furnished an excellent meal for everybody in camp. This evening at sunset the view across the broad river, from our camp where the two rivers joined, was very lovely; and for the first time we had an open space in front of and above us, so that after nightfall the stars, and the great waxing moon, were glorious overhead, and against the rocks in midstream the broken water gleamed like tossing silver.

The huge catfish which the men had caught was over three feet and a half long, with the unusual enormous head, out of all proportion to the body, and the enormous mouth, out of all proportion to the head. Such fish, although their teeth are small, swallow very large prey. This one contained the nearly digested remains of a monkey. Probably the monkey had been seized while drinking from the end of a branch; and once engulfed in that yawning cavern there was no escape. We Americans were astounded at the idea of a catfish making prey of a monkey; but our Brazilian friends told us that in the lower Madeira and the part of the Amazon near its mouth there is a still more gigantic catfish which in similar fashion occasionally makes prey of man. This is a grayish-white fish over nine feet long, with the usual disproportionately large head and gaping mouth, with a circle of small teeth; for the engulfing mouth itself is the danger, not the teeth. It is called the piraiba—pronounced in four syllables. While stationed at the small city of Itacoatiara, on the Amazon, at the mouth of the Madeira, the doctor had seen one of these monsters which had been killed by the two men it had attacked. They were fishing in a canoe when it rose from the bottom—for it is a ground fish—and raising itself half out of the water lunged over the edge of the canoe at them, with open mouth. They killed it with their *falcóns*, as machetes are called in Brazil. It was taken round the city in triumph in an ox-cart; the doctor saw it and said it was three metres long. He said that swimmers feared it even more than the big cayman, because they could see the latter, whereas the former lay hid at the bottom of the water. Colonel Rondon said that in many villages where he had been on the lower Madeira the people had built stockaded enclosures in the water in which they bathed, not venturing to swim in the open water for fear of the piraiba and the big cayman.

Next day, April 8, we made five kilometres only, as there was a succession of rapids. We had to carry the loads past two of them but ran the canoes without difficulty, for on the west side were long canals of swift water through the forest. The river had been higher but was still very high, and the current raced round the many islands that at this point divided the channel. At four we made camp at the head of another stretch of rapids, over which the Canadian canoes would have danced without shipping a teaspoonful of water, but which our dugouts could only run empty. Cherrie killed three monkeys and Lyra caught two big piranhas, so that we were again all of us well provided with dinner and breakfast. When a number of men, doing hard work, are most of the time on half-rations, they grow to take a lively interest in any reasonably full meal that does arrive.

On the 10th we repeated the proceedings: a short quick run; a few hundred metres' portage, occupying, however, at least a couple of hours; again a few minutes' run; again other rapids. We again made less than five kilometres; in the two days we had been descending nearly a metre for every kilometre we made in advance; and it hardly seemed as if this state of things could last, for the aneroid showed that we were getting very low down. How I longed for a big Maine birchbark, such as that in which I once went down the Mattawamkeag at high water! It would have slipped down these rapids as a girl trips through a country-dance. But our loaded dugouts would have shoved their noses under every curl. The country was lovely. The wide river, now in one channel, now in several channels, wound among hills; the shower-freshened forest glistened in the sunlight; the many kinds of beautiful palm fronds and the huge pacova-leaves stamped the peculiar look of the tropics on the whole landscape—it was like passing by water through a gigantic botanical garden. In the afternoon we got an elderly toucan, a piranha, and a reasonably edible side-necked river-turtle; so we had fresh meat again. We slept as usual in earshot of rapids. We had been out six weeks and almost all the time we had been engaged in wearily working our way down and past rapid after rapid. Rapids are by far the most dangerous enemies of explorers and travelers who journey along these rivers.

Next day was a repetition of the same work. All the morning was spent in getting the loads to the foot of the rapids at the head of which we were encamped, down which the canoes were run empty. Then for thirty or forty minutes we ran down the swift, twisting river, the two lashed canoes almost coming to grief at one spot where a swirl of the current threw them against some trees on a small submerged island. Then we came to another set of rapids, carried the baggage down past them, and made camp long after dark in the rain—a good exercise in patience for those of us who were still suffering somewhat from fever. No one was in really buoyant health. For some weeks we had been sharing part of the contents of our boxes with the camaradas; but our food was not very satisfying to them. They needed quantity and the mainstay of each of their meals was a mass of palmitas; but on this day they had no time to cut down palms. We finally decided to run these rapids with the empty canoes, and they came down in safety. On such a trip it is highly undesirable to take any save necessary risks, for the consequences of disaster are too serious; and yet if no risks are taken the progress is so slow that disaster comes anyhow; and it is necessary perpetually to vary the term of the perpetual working compromise between rashness and overcaution. This night we had a very good fish to eat, a big silvery fellow called a pescada, of a kind we had not caught before.

One day Trigueiro failed to embark with the rest of us, and we had to camp where we were next day to find him. Easter Sunday we spent in the fashion with which we were altogether too familiar. We only ran in a clear course for ten minutes all told, and spent eight hours in portaging the loads past rapids down which the canoes were run; the balsa was almost swamped. This day we caught twenty-eight big fish, mostly piranhas, and everybody had all he could eat for dinner and for breakfast the following morning.

The forenoon of the following day was a repetition of this wearisome work; but late in the afternoon the river began to run in long and quiet reaches. We made fifteen kilometres, and for the first time in several weeks camped where we did not hear the rapids. The silence was soothing and restful. The following day, April 14, we made a good run of some thirty-two kilometres. We passed a little river which entered on our left. We ran two or three light rapids and portaged the loads by another. The river ran in long and unusually tranquil stretches. In the morning when we started the view was lovely. There was a mist, and for a couple of miles the great river, broad and quiet, ran between the high walls of tropical forest, the tops of the giant trees showing dim through the haze. Different members of the party caught many fish, and shot a monkey and a couple of jacu-tinga—birds kin to a turkey but the size of a fowl—so we again had a camp of plenty. The dry season was approaching, but there were still heavy, drenching rains. On this day the men found some new nuts of which they liked the taste;

but the nuts proved unwholesome and half of the men were very sick and unable to work the following day. In the balsa only two were left fit to do anything, and Kermit plied a paddle all day long.

Accordingly, it was a rather sorry crew that embarked the following morning, April 15. But it turned out a red-letter day. The day before, we had come across cuttings, a year old, which were probably but not certainly made by pioneer rubber men. But on this day—during which we made twenty-five kilometres—after running two hours and a half we found on the left bank a board on a post with the initials J. A., to show the farthest-up point which a rubber man had reached and claimed as his own. An hour farther down we came on a newly built house in a little planted clearing; and we cheered heartily. No one was at home, but the house of palm thatch was clean and cool. A couple of dogs were on the watch, and the belongings showed that a man, and a woman, and a child lived there and had only just left. Another hour brought us to a similar house where dwelt an old black man who showed the innate courtesy of the Brazilian peasant. We came on these rubber men and their houses in about latitude ten degrees twenty-four minutes.

In mid-afternoon we stopped at another clean, cool, picturesque house of palm thatch. The inhabitants all fled at our approach, fearing an Indian raid; for they were absolutely unprepared to have any one come from the unknown regions upstream. They returned and were most hospitable and communicative; and we spent the night there. Said Antonio Correa to Kermit: "It seems like a dream to be in a house again and hear the voices of men and women, instead of being among those mountains and rapids." The river was known to them as the Castanho and was the main affluent, or rather the left or western branch, of the Aripuanan; the Castanho is a name used by the rubber-gatherers only; it is unknown to the geographers. We were, according to our informants, about fifteen days' journey from the confluence of the two rivers; but there were many rubber men along the banks, some of whom had become permanent settlers. We had come over three hundred kilometres in forty-eight days, over absolutely unknown ground; we had seen no human being, although we had twice heard Indians. Six weeks had been spent in steadily slogging our way down the interminable series of rapids. It was astonishing before, when we were on a river of about the size of the upper Rhine or Elbe, to realize that no geographer had any idea of its existence. But, after all, no civilized man of any grade had ever been on it. Here, however, was a river with people dwelling along the banks, some of whom had lived in the neighborhood for eight or ten years; and yet on no standard map was there a hint of the river's existence. We were putting on the map a river, running through between five and six degrees of latitude—of between seven and eight if, as should properly be done, the lower Aripuanan is included as part of it—of which no geographer, in any map published in Europe or the United States or Brazil, had even admitted the possibility of the existence; for the place actually occupied by it was filled, on the maps, by other—imaginary—streams or by mountain ranges. Before we started, the Amazonas Boundary Commission had come up the lower Aripuanan and then the eastern branch, or upper Aripuanan, to eight degrees forty-eight minutes, following the course which for a couple of decades had been followed by the rubber men, but not going as high. An employee, either of this commission or of one of the big rubber men, had been up the Castanho, which is easy of ascent in its lower course, to about the same latitude, not going nearly as high as the rubber men had gone; this we found out as we ourselves were descending the lower Castanho. The lower main stream, and the lower portion of its main affluent, the Castanho, had been commercial highways for rubber men and settlers for nearly two decades, and, as we speedily found, were as easy to traverse as the upper stream, which we had just come down, was difficult to traverse; but the governmental and scientific authorities, native and foreign, remained in complete ignorance; and the rubber men themselves had not the slightest idea of the headwaters, which were in country never hitherto traversed by civilized man. Evidently the Castanho was, in length at least, substantially equal, and probably superior, to the upper Aripuanan; it now seemed even more likely that the Ananás was the headwaters of the main stream than of the Cardozo.[3] For the first time this great river, the greatest affluent of the Madeira, was to be put on the map; and the understanding of its real position and real relationship, and the clearing up of the complex problem of the sources of all these lower right-hand affluents of the Madeira, were rendered possible by the seven weeks of hard and dangerous labor we had spent in going down an absolutely unknown river, through an absolutely unknown wilderness. At this stage of the growth of world geography I esteemed it a great piece of good fortune to be able to take part in such a feat—a feat which represented the capping of the pyramid which during the previous seven years had been built by the labor of the Brazilian Telegraphic Commission.

We had passed the period when there was a chance of peril, of disaster, to the whole expedition. There might be risk ahead to individuals, and some difficulties and annoyances for all of us; but there was no longer the least likelihood of any disaster to the expedition as a whole. We now no longer had to face continual anxiety, the need of constant economy with food, the duty of labor with no end in sight, and bitter uncertainty as to the future.

It was time to get out. The wearing work, under very unhealthy conditions, was beginning to tell on every one. Half of the camaradas had been down with fever and were much weakened; only a few of them retained their original physical and moral strength. Cherrie and Kermit had recovered; but both Kermit and Lyra still had bad sores on their legs from the bruises received in the water work. I was in worse shape. The after-effects of the fever still hung on; and the leg which had been hurt while working with the rapids with the sunken canoe had taken a turn for the bad and developed an abscess. The good doctor, to whose unwearied care and kindness I owe much, had cut it open and inserted a drainage-tube; an added charm being given the operation and the subsequent dressings by the enthusiasm with which the piums and boroshudas took part therein. I could hardly hobble and was pretty well laid up. But "there aren't no 'stop, conductor,' while a battery's changing ground." No man has any business to go on such a trip as ours unless he will refuse to jeopardize the welfare of his associates by any delay caused by a weakness or ailment of his. It is his duty to go forward, if necessary on all fours, until he drops. Fortunately, I was put to no such test. I remained in good shape until we had passed the last of the rapids of the chasms. When my serious trouble came we had only canoe-riding ahead of us. It is not ideal for a sick man to spend the hottest hours of the day stretched on the boxes in the bottom of a small open dugout, under the well-nigh intolerable heat of the torrid sun of the mid-tropics, varied by blinding, drenching downpours of rain; but I could not be sufficiently grateful for the chance. Kermit and Cherrie took care of me as if they had been trained nurses; and Colonel Rondon and Lyra were no less thoughtful.

The north was calling strongly to the three men of the north—Rocky Dell Farm to Cherrie, Sagamore Hill to me; and to Kermit the call was stronger still. After nightfall we could now see the Dipper well above the horizon—upside down, with the two pointers pointing to a north star below the world's rim; but the Dipper, with all its stars. In our home country spring had now come, the wonderful northern spring of long glorious days, of brooding twilights, of cool delightful nights. Robin and bluebird, meadow-lark and song-sparrow, were singing in the mornings at home; the maple buds were red; wind-flowers and bloodroot were blooming while the last patches of snow still lingered; the rapture of the hermit-thrush in Vermont, the serene golden melody of the wood-thrush on Long Island, would be heard before we were there to listen. Each man to his home, and to his true love! Each was longing for the homely things that were so dear to him, for the home people who were dearer still, and for the one who was dearest of all.

# PART 2

# WILDERNESS PRESERVATION

# 6

# A National Park Service

THE AMERICAN PEOPLE HAVE BEEN GRADUALLY AWAKENING during the past few years to the idea of the city park as not merely an adornment, but an instrument of social service to the community. We are coming more and more to realize that life is better worth living in those cities which have relatively large park areas effectively developed not only as beauty spots but as recreation centers and playgrounds for all classes in the community. But we have not yet made effective application of the same idea to our National parks. It is true that we have an ample measure of them in area, but we are not yet sure as a people just what we want them for; and we have as yet given them no efficient and intelligent administration. There are in the United States thirteen National parks, embracing over four and half million acres. At present, as the Secretary of the Interior has pointed out in his annual report, each of these parks is a separate and distinct unit for administrative purposes. Special appropriations are made for each park, and the employment of a common supervising and directing force is impossible. As the President of the American Civic Association said in his address at the Association's recent convention, "Nowhere in official Washington can an inquirer find an office of the National parks, or a desk devoted solely to their management. By passing around through their departments, and consulting clerks who have taken on the extra work of doing what they can for the Nation's playgrounds, it is possible to come at a little information." A bill is before Congress for the creation of a Bureau of National Parks, the head of which shall have the supervision, management, and control of all the National parks and National monuments in the country, and shall have the duty of developing these areas so that they shall be the most efficient agencies possible for promoting public recreation and public health through their use and enjoyment by the people. We have a single amendment to propose to the bill. The new bureau should be called the National Park Service, in conformity with the custom already established in naming the Forest Service. The establishment of the National Park service is justified by considerations of good administration, of the value of natural beauty as a National asset, and of the effectiveness of outdoor life and recreation in the production of good citizenship.

# 7

# John Muir

*An Appreciation*

OUR GREATEST NATURE-LOVER AND NATURE-WRITER, THE man who has done most in securing for the American people the incalculable benefit of appreciation of the wild nature in his own land, is John Burroughs. Second only to John Burroughs, and in some respects ahead even of John Burroughs, was John Muir. Ordinarily, the man who loves the woods and the mountains, the trees, the flowers, and the wild things, has in him some indefinable quality of charm which appeals even to those sons of civilization who care for little outside of paved streets and brick walls. John Muir was a fine illustration of this rule. He was by birth a Scotchman—a tall and spare man, with the poise and ease natural to him who has lived much alone under conditions of labor and hazard. His was a dauntless soul, and also one brimming over with friendliness and kindliness.

He was emphatically a good citizen. Not only are his books delightful, not only is he the author to whom all men turn when they think of the Sierras and Northern glaciers, and the giant trees of the California slope, but he was also—what few nature-lovers are—a man able to influence contemporary thought and action on the subjects to which he had devoted his life. He was a great factor in influencing the thought of California and the thought of the entire country so as to secure the preservation of those great natural phenomena—wonderful canyons, giant trees, slopes of flower-spangled hillsides—which make California a veritable Garden of the Lord.

It was my good fortune to know John Muir. He had written me, even before I met him personally, expressing his regret that when Emerson came to see the Yosemite, his (Emerson's) friends would not allow him to accept John Muir's invitation to spend two or three days camping with him, so as to see the giant grandeur of the place under surroundings more congenial than those of a hotel piazza or a seat on a coach. I had answered him that if ever I got in his neighborhood I should claim from him the treatment that he had wished to accord Emerson. Later, when as President I visited the Yosemite, John Muir fulfilled the promise he had at that time made to me. He met me with a couple of pack-mules, as well as with riding mules for himself and myself, and a first-class packer and cook, and I spent a delightful three days and two nights with him.

The first night we camped in a grove of giant sequoias. It was clear weather, and we lay in the open, the enormous cinnamon-colored trunks rising about us like the columns of a vaster and more beautiful cathedral then was ever conceived by any human architect. One incident surprised me not a little. Some thrushes—I think they were Western hermit-thrushes—were singing beautifully in the solemn evening stillness. I asked some question concerning them of John Muir, and to my surprise found that he had not been listening to them and knew nothing about them. Once or twice I had been off with John Burroughs, and had found that, although he was so much older than I was, his ear and his eye were infinitely better as regards the sights and sounds of wild life, or at least of the smaller wild life, and I was accustomed unhesitatingly to refer to him regarding any bird note that puzzled me. But John Muir, I found, was not interested in the small things of nature unless they were unusually conspicuous. Mountains, cliffs, trees, appealed to him tremendously, but birds did not unless they possessed some very peculiar and interesting as well as conspicuous traits, as in the case of the water-ouzel. In the same way, he knew nothing of the wood-mice; but the more conspicuous beasts, such as bear and deer, for example, he could tell much about.

All next day we travelled through the forest. Then a snowstorm came on, and at night we camped on the edge of the Yosemite, under the branches of a magnificent silver fir, and very warm and comfortable we were, and a very good dinner we had before we rolled up in our tarpaulins and blankets for the night. The following day we went down into the Yosemite and through the valley, camping in the bottom among the timber.

There was a delightful innocence and good-will about the man, and an utter inability to imagine that any one could either take or give offense. Of this I had an amusing illustration just before we parted. We were saying good-by, when his expression suddenly changed, and he remarked that he had totally forgotten something. He was intending to go to the Old World with a great tree-lover and tree-expert from the Eastern States who possessed a somewhat crotchety temper. He informed me that his friend had written him, asking him to get from me personal letters to the Russian Czar and the Chinese Emperor; and when I explained to him that I could not give personal letters to foreign potentates, he said: "Oh, well, read the letter yourself, and that will explain just what I want." Accordingly he thrust the letter on me. It contained not only the request which he had mentioned, but also a delicious preface, which, with the request, ran somewhat as follows:

"I hear Roosevelt is coming out to see you. He takes a sloppy, unintelligent interest in forests, although he is altogether too much under the influence of that creature Pinchot, and you had better get from him letters to the Czar of Russia and the Emperor of China, so that we may have better opportunity to examine the forests and trees of the Old World."

Of course I laughed heartily as I read the letter, and said: "John, do you remember exactly the words in which this letter was couched?" Whereupon a look of startled surprise came over his face, and he said: "Good gracious! There was something unpleasant about you in it, wasn't there? I had forgotten. Give me the letter back."

So I gave him back the letter, telling him that I appreciated it far more than if it had not contained the phrases he had forgotten, and that while I could not give him and his companion letters to the two rulers in question, I would give him letters to our ambassadors, which would bring about the same result.

John Muir talked even better than he wrote. His greatest influence was always upon those who were brought into personal contact with him. But he wrote well, and while his books have not the peculiar charm that a very, very few other writers on similar subjects have had, they will nevertheless last long. Our generation owes much to John Muir.

# 8

# Wilderness Reserves

## *The Yellowstone Park*

THE MOST STRIKING AND MELANCHOLY FEATURE IN CONNECTION with American big game is the rapidity with which it has vanished. When, just before the outbreak of the Revolutionary War, the rifle-bearing hunters of the backwoods first penetrated the great forests west of the Alleghanies, deer, elk, black bear, and even buffalo, swarmed in what are now the States of Kentucky and Tennessee, and the country north of the Ohio was a great and almost virgin hunting-ground. From that day to this the shrinkage has gone on, only partially checked here and there, and never arrested as a whole. As a matter of historical accuracy, however, it is well to bear in mind that many writers, in lamenting this extinction of the game, have from time to time anticipated or overstated the facts. Thus as good an author as Colonel Richard Irving Dodge spoke of the buffalo as practically extinct, while the great Northern herd still existed in countless thousands. As early as 1880 sporting authorities spoke not only of the buffalo but of the elk, deer, and antelope as no longer to be found in plenty; and recently one of the greatest of living hunters has stated that it is no longer possible to find any American wapiti bearing heads comparable with the red deer of Hungary. As a matter of fact, in the early eighties there were still large regions where every species of game that had ever been known within historic times on our continent was still to be found as plentifully as ever. In the early nineties there were still big tracts of wilderness in which this was true of all game except the buffalo; for instance, it was true of the elk in portions of northwestern Wyoming, of the blacktail in northwestern Colorado, of the whitetail here and there in Indian Territory, and of the antelope in parts of New Mexico. Even at the present day there are smaller, but still considerable, regions where these four animals are yet found in abundance; and I have seen antlers of wapiti shot since 1900 far surpassing any of which there is record from Hungary. In New England and New York, as well as New Brunswick and Nova Scotia, the whitetail deer is more plentiful than it was thirty years ago, and in Maine (and to an even greater extent in New Brunswick) the moose, and here and there the caribou, have, on the whole, increased during the same period. There is yet ample opportunity for the big-game hunter in the United States, Canada, and Alaska.

While it is necessary to give this word of warning to those who, in praising time past, always forget the opportunities of the present, it is a thousandfold more necessary to remember that these opportunities are, nevertheless, vanishing; and if we are a sensible people, we will make it our business to see that the process of extinction is arrested. At the present moment the great herds of caribou are being butchered, as in the past the great herds of bison and wapiti have been butchered. Every believer in manliness and therefore in manly sport, and every lover of nature, every man who appreciates the majesty and beauty of the wilderness and of wild life, should strike hands with the farsighted men who wish to preserve our material resources, in the effort to keep our forests and our game beasts, game-birds, and game-fish—indeed, all the living creatures of prairie and woodland and seashore—from wanton destruction.

Above all, we should realize that the effort toward this end is essentially a democratic movement. It is entirely in our power as a nation to preserve large tracts of wilderness, which are valueless for agricultural purposes and unfit for settlement, as playgrounds for rich and poor alike, and to preserve the game so that it shall continue to exist for the benefit of all lovers of nature, and to give reasonable opportunities for the exercise of the skill of the hunter, whether he is or is not a man of means. But this end can only be achieved by wise laws and by a resolute enforcement of the laws.

Lack of such legislation and administration will result in harm to all of us, but most of all in harm to the nature-lover who does not possess vast wealth. Already there have sprung up here and there through the country, as in New Hampshire and the Adirondacks, large private preserves. These preserves often serve a useful purpose, and should be encouraged within reasonable limits; but it would be a misfortune if they increased beyond a certain extent or if they took the place of great tracts of wild land, which continue as such either because of their very nature, or because of the protection of the State exerted in the form of making them State or national parks or reserves. It is foolish to regard proper game-laws as undemocratic, unrepublican. On the contrary, they are essentially in the interests of the people as a whole, because it is only through their enactment and enforcement that the people as a whole can preserve the game and can prevent its becoming purely the property of the rich, who are able to create and maintain extensive private preserves. The wealthy man can get hunting anyhow, but the man of small means is dependent solely upon wise and well-executed game-laws for his enjoyment of the sturdy pleasure of the chase. In Maine, in Vermont, in the Adirondacks, even in parts of Massachusetts and on Long Island, people have waked up to this fact, particularly so far as the common whitetail deer is concerned, and in Maine also as regards the moose and caribou. The effect is shown in the increase in these animals. Such game protection results, in the first place, in securing to the people who live in the neighborhood permanent opportunities for hunting; and in the next place, it provides no small source of wealth to the locality because of the visitors which it attracts. A deer wild in the woods is worth to the people of the neighborhood many times the value of its carcass, because of the way it attracts sportsmen, who give employment and leave money behind them.

True sportsmen, worthy of the name, men who shoot only in season and in moderation, do no harm whatever to game. The most objectionable of all game-destroyers is, of course, the kind of game-butcher who simply kills for the sake of the record of slaughter, who leaves deer and ducks and prairie-chickens to rot after he has slain them. Such a man is wholly obnoxious; and, indeed, so is any man who shoots for the purpose of establishing a record of the amount of game killed. To my mind this is one very unfortunate feature of what is otherwise the admirably sportsmanlike English spirit in these matters. The custom of shooting great bags of deer, grouse, partridges, and pheasants, the keen rivalry in making such bags, and their publication in sporting journals, are symptoms of a spirit which is most unhealthy from every standpoint. It is to be earnestly hoped that every American hunting or fishing club will strive to inculcate among its own members, and in the minds of the general public, that anything like an excessive bag, any destruction for the sake of making a record, is to be severely reprobated.

But, after all, this kind of perverted sportsman, unworthy though he be, is not the chief actor in the destruction of our game. The professional skin or market hunter is the real offender. Yet he is of all others the man who would ultimately be most benefited by the preservation of the game. The frontier settler, in a thoroughly wild country, is certain to kill game for his own use. As long as he does no more than this, it is hard to blame him; although if he is awake to his own interests he will soon realize that to him, too, the live deer is worth far more than the dead deer, because of the way in which it brings money into the wilderness. The professional market-hunter who kills game for the hide or for the feathers or for the meat or to sell antlers and other trophies; the marketmen who put game in cold storage; and the rich people, who are content to buy what they have not the skill to get by their own exertions—these are the men who are the real enemies of game. Where there is no law which checks the market-hunters, the inevitable result of their butchery is that the game is completely destroyed, and with it their own means of livelihood. If, on the other hand, they were willing to preserve it, they could make much more money by acting as guides. In northwestern Colorado, at the present moment, there are still blacktail deer in abundance, and some elk are left. Colorado has fairly good game-laws, but they are indifferently enforced. The country in which the game is found can probably never support any but a very sparse population, and a large portion of the summer range is practically useless for settlement. If the people of Colorado generally, and above all the people of the counties in which the game is located, would resolutely co-operate with those of their own number who are already alive to the importance of preserving the game, it could, without difficulty, be kept always as abundant as it now is, and this beautiful region would be a permanent health resort and playground for the people of a large part of the Union. Such action would be a benefit to every one, but it would be a benefit most of all to the people of the immediate locality.

The practical common sense of the American people has been in no way made more evident during the last few years than by the creation and use of a series of large land reserves—situated for the most part on the great plains and among the mountains of the West—intended to keep the forests from destruction, and therefore to conserve the water-supply. These reserves are, and should be, created primarily for economic purposes. The semiarid regions can only support a reasonable population under conditions of the strictest economy and wisdom in the use of the water-supply, and in addition to their other economic uses the forests are indispensably necessary for the preservation of the water-supply and for rendering possible its useful distribution throughout the proper seasons. In addition, however, to this economic use of the wilderness, selected portions of it have been kept here and there in a state of nature, not merely for the sake of preserving the forests and the water, but for the sake of preserving all its beauties and wonders unspoiled by greedy and shortsighted vandalism. What has been actually accomplished in the Yellowstone Park affords the best possible object-lesson as to the desirability and practicability of establishing such wilderness reserves. This reserve is a natural breeding-ground and nursery for those stately and beautiful haunters of the wilds which have now vanished from so many of the great forests, the vast lonely plains, and the high mountain ranges, where they once abounded.

On April 8, 1903, John Burroughs and I reached the Yellowstone Park, and were met by Major John Pitcher of the regular army, the superintendent of the Park. The major and I forthwith took horses; he telling me that he could show me a good deal of game while riding up to his house at the Mammoth Hot Springs. Hardly had we left the little town of Gardiner and gotten within the limits of the Park before we saw a prongbuck. There was a band of at least a hundred feeding some distance from the road. We rode leisurely toward them. They were tame compared to their kindred in unprotected places; that is, it was easy to ride within fair rifle-range of them; and though they were not familiar in the sense that we afterward found the bighorn and the deer to be familiar, it was extraordinary to find them showing such familiarity almost literally in the streets of a frontier town. It spoke volumes for the good sense and law-abiding spirit of the people of the town. During the two hours following my entry into the Park we rode around the plains and lower slopes of the foot-hills in the neighborhood of the mouth of the Gardiner and we saw several hundred—probably a thousand all told—of these antelopes. Major Pitcher informed me that all the pronghorns in the Park wintered in this neighborhood. Toward the end of April or the first of May they migrate back to their summering homes in the open valleys along the Yellowstone and in the plains south of the Golden Gate. While migrating they go over the mountains and through forests if occasion demands. Although there are plenty of coyotes in the Park, there are no big wolves, and save for very infrequent poachers, the only enemy of the antelope, as indeed the only enemy of all the game, is the cougar.

Cougars, known in the Park, as elsewhere through the West, as "mountain-lions," are plentiful, having increased in numbers of recent years. Except in the neighborhood of the Gardiner River—that is within a few miles of Mammoth Hot Springs—I found them feeding on elk, which in the Park far outnumber all other game put together, being so numerous that the ravages of the cougars are of no real damage to the herds. But in the neighborhood of the Mammoth Hot Springs the cougars are noxious because of the antelope, mountain-sheep, and deer which they kill; and the superintendent has imported some hounds with which to hunt them. These hounds are managed by Buffalo Jones, a famous old plainsman, who is now in the Park taking care of the buffalo. On this first day of my visit to the Park I came across the carcass of a deer and of an antelope which the cougars had killed. On the great plains cougars rarely get antelope, but here the country is broken so that the big cats can make their stalks under favorable circumstances. To deer and mountain-sheep the cougar is a most dangerous enemy—much more so than the wolf.

The antelope we saw were usually in bands of from twenty to one hundred and fifty, and they travelled strung out almost in single file, though those in the rear would sometimes bunch up. I did not try to stalk them, but got as near them as I could on horseback. The closest approach I was able to make was to within about eighty yards of two which were by themselves—I think, a doe and a last year's fawn. As I was riding up to them, although they looked suspiciously at me, one actually lay down. When I was passing them at about eighty yards' distance the big one became nervous, gave a sudden jump, and away the two went at full speed.

Why the prongbucks were so comparatively shy I do not know, for right on the ground with them we came upon deer, and, in the immediate neighborhood, mountain-sheep, which were absurdly tame. The mountain-sheep were nineteen in number—for the most part does and yearlings with a couple of three-year-old rams—but not a single big fellow, for the big fellows at this season are off by themselves, singly or in little bunches, high up in the mountains. The band I saw was tame to a degree matched by but a few domestic animals.

They were feeding on the brink of a steep washout at the upper edge of one of the benches on the mountainside just below where the abrupt slope began. They were alongside a little gully with sheer walls. I rode my horse to within forty yards of them, one of them occasionally looking up and at once continuing to feed. Then they moved slowly off, and leisurely crossed the gully to the other side. I dismounted, walked around the head of the gully, and moving cautiously, but in plain sight, came closer and closer until I was within twenty yards, when I sat down on a stone and spent certainly twenty minutes looking at them. They paid hardly any attention to my presence—certainly no more than well-treated domestic creatures would pay. One of the rams rose on his hind legs, leaning his fore hoofs against a little pine-tree, and browsed the ends of the budding branches. The others grazed on the short grass and herbage or lay down and rested—two of the yearlings several times playfully butting at one another. Now and then one would glance in my direction, without the slightest sign of fear—

barely even of curiosity. I have no question whatever but that with a little patience this particular band could be made to feed out of a man's hand. Major Pitcher intends during the coming winter to feed them alfalfa—for game animals of several kinds have become so plentiful in the neighborhood of the Hot Springs and the major has grown so interested in them that he wishes to do something toward feeding them during the severe weather. After I had looked at the sheep to my heart's content, I walked back to my horse, my departure arousing as little interest as my advent.

Soon after leaving them we began to come across blacktail deer, singly, in twos and threes, and in small bunches of a dozen or so. They were almost as tame as the mountain-sheep, but not quite; that is, they always looked alertly at me, and though if I stayed still they would graze, they kept a watch over my movements and usually moved slowly off when I got within less than forty yards of them. Up to that distance, whether on foot or on horseback, they paid but little heed to me, and on several occasions they allowed me to come much closer. Like the bighorn, the blacktails at this time were grazing, not browsing; but I occasionally saw them nibble some willow buds. During the winter they had been browsing. As we got close to the Hot Springs we came across several whitetail in an open, marshy meadow. They were not quite as tame as the blacktail, although without any difficulty I walked up to within fifty yards of them. Handsome though the blacktail is, the whitetail is the most beautiful of all deer when in motion, because of the springy, bounding grace of its trot and canter, and the way it carries its head and white flag aloft.

Before reaching the Mammoth Hot Springs we also saw a number of ducks in the little pools and on the Gardiner. Some of them were rather shy. Others—probably those which, as Major Pitcher informed me, had spent the winter there—were as tame as barnyard fowls.

Just before reaching the post the major took me into the big field where Buffalo Jones had some Texas and Flathead Lake buffalo—bulls and cows—which he was tending with solicitous care. The original stock of buffalo in the Park have now been reduced to fifteen or twenty individuals, and their blood is being recruited by the addition of buffalo purchased out of the Flathead Lake and

Texas Panhandle herds. The buffalo were at first put within a wire fence, which, when it was built, was found to have included both blacktail and whitetail deer. A bull elk was also put in with them at one time, he having met with some accident which made the major and Buffalo Jones bring him in to doctor him. When he recovered his health he became very cross. Not only would he attack men, but also buffalo, even the old and surly master-bull, thumping them savagely with his antlers if they did anything to which he objected. The buffalo are now breeding well.

When I reached the post and dismounted at the major's house, I supposed my experiences with wild beasts were ended for the day; but this was an error. The quarters of the officers and men and the various hotel buildings, stables, residence of the civilian officials, etc., almost completely surround the big parade-ground at the post, near the middle of which stands the flagpole, while the gun used for morning and evening salutes is well off to one side. There are large gaps between some of the buildings, and Major Pitcher informed me that throughout the winter he had been leaving alfalfa on the parade-grounds, and that numbers of blacktail deer had been in the habit of visiting it every day, sometimes as many as seventy being on the parade-ground at once. As springtime came on, the numbers diminished. However, in mid-afternoon, while I was writing in my room in Major Pitcher's house, on looking out of the window I saw five deer on the parade-ground. They were as tame as so many Alderney cows, and when I walked out I got within twenty yards of them without any difficulty. It was most amusing to see them as the time approached for the sunset gun to be fired. The notes of the trumpeter attracted their attention at once. They all looked at him eagerly. One of them resumed feeding, and paid no attention whatever either to the bugle, the gun, or the flag. The other four, however, watched the preparations for firing the gun with an intent gaze, and at the sound of the report gave two or three jumps; then instantly wheeling, looked up at the flag as it came down. This they seemed to regard as something rather more suspicious than the gun, and they remained very much on the alert until the ceremony was over. Once it was finished, they resumed feeding as if nothing had happened. Before it was dark they trotted away from the parade-ground back to the mountains.

The next day we rode off to the Yellowstone River, camping some miles below Cottonwood Creek. It was a very pleasant camp. Major Pitcher, an old friend, had a first-class pack-train, so that we were as comfortable as possible, and on such a trip there could be no pleasanter or more interesting companion than John Burroughs—"Oom John," as we soon grew to call him. Where our tents were pitched the bottom of the valley was narrow, the mountains rising steep and cliff-broken on either side. There were quite a number of blacktail in the valley, which were tame and unsuspicious, although not nearly as much so as those in the immediate neighborhood of the Mammoth Hot Springs. One mid-afternoon three of them swam across the river a hundred yards above our camp. But the characteristic animals of the region were the elk—the wapiti. They were certainly more numerous than when I was last through the Park twelve years before.

In the summer the elk spread all over the interior of the Park. As winter approaches they divide, some going north and others south. The southern bands, which, at a guess, may possibly include ten thousand individuals, winter out of the Park, for the most part in Jackson's Hole—though of course here and there within the limits of the Park a few elk may spend both winter and summer in an unusually favorable location. It was the members of the northern band that I met. During the winter-time they are nearly stationary, each band staying within a very few miles of the same place, and from their size and the open nature of their habitat it is almost as easy to count them as if they were cattle. From a spur of Bison Peak one day, Major Pitcher, the guide Elwood Hofer, John Burroughs, and I spent about four hours with the glasses counting and estimating the different herds within sight. After most careful work and cautious reduction of estimates in each case to the minimum the truth would permit, we reckoned three thousand head of elk, all lying or feeding and all in sight at the same time. An estimate of some fifteen thousand for the number of elk in these northern bands cannot be far wrong. These bands do not go out of the Park at all, but winter just within its northern boundary. At the time when we saw them, the snow had vanished from the bottoms of the valleys and the lower slopes of the mountains, but remained as continuous sheets farther up their sides. The elk were for the most part found up on the snow slopes, occasionally singly or in small gangs—more often in bands of from fifty to a couple of hundred. The larger bulls were highest up the mountains and generally in small troops by themselves, although occasionally one or two would be found associating with a big herd of cows, yearlings, and two-year-olds. Many of the bulls had shed their antlers; many had not. During the winter the elk had evidently done much browsing, but at this time they were grazing almost exclusively, and seemed by preference to seek out the patches of old grass which were last left bare by the retreating snow. The bands moved about very little, and if one were seen one day it was generally possible to find it within a few hundred yards of the same spot the next day, and certainly not more than a mile or two off. There were severe frosts at night, and occasionally light flurries of snow; but the hardy beasts evidently cared nothing for any but heavy storms, and seemed to prefer to lie in the snow rather than upon the open ground. They fed at irregular hours throughout the day, just like cattle; one band might be lying down while another was feeding. While travelling they usually went almost in single file. Evidently the winter had weakened them, and they were not in condition for running; for on the one or two occasions when I wanted to see them close up I ran right into them on horseback, both on level plains and going uphill along the sides of rather steep mountains. One band in particular I practically rounded up for John Burroughs, finally getting them to stand in a huddle while he and I sat on our horses less than fifty yards off. After they had run a little distance they opened their mouths wide and showed evident signs of distress.

We came across a good many carcasses. Two, a bull and a cow, had died from scab. Over half the remainder had evidently perished from cold or starvation. The others, including a bull, three cows, and a score of yearlings, had been killed by cougars. In the Park the cougar is at present their only animal foe. The cougars were preying on nothing but elk in the Yellowstone Valley, and kept hanging about the neighborhood of the big bands. Evidently they usually selected some outlying yearling, stalked it as it lay or as it fed, and seized it by the head and throat. The bull which they killed was in a little open valley by himself, many miles from any other elk. The cougar which killed it, judging from its tracks, was a big male. As the elk were evidently rather too numerous for the feed, I do not think the cougars were doing any damage.

Coyotes are plentiful, but the elk evidently have no dread of them. One day I crawled up to within fifty yards of a band of elk lying down. A coyote was walking about among them, and beyond an occasional look they paid no heed to him. He did not venture to go within fifteen or twenty paces of any one of them. In fact, except the cougar, I saw but one living thing attempt to molest the elk. This was a golden eagle. We saw several of these great birds. On one occasion we had ridden out to the foot of a sloping mountainside, dotted over with bands and strings of elk amounting in the aggregate probably to a thousand head. Most of the bands were above the snow-line—some appearing away back toward the ridge crests, and looking as small as mice. There was one band well below the snow-line, and toward this we rode. While the elk were not shy or wary, in the sense that a hunter would use the words, they were by no means as familiar as the deer; and this particular band of elk, some twenty or thirty in all, watched us with interest as we approached. When we were still half a mile off they suddenly started to run toward us, evidently frightened by something. They ran quartering, and when about four hundred yards away we saw that an eagle was after them. Soon it swooped, and a yearling in the rear, weakly, and probably frightened by the swoop, turned a complete somersault, and when it recovered its feet stood still. The great bird followed the rest of the band across a little ridge, beyond which they disappeared. Then it returned, soaring high in the heavens, and, after two or three wide circles, swooped down at the solitary yearling, its legs hanging down. We halted at two hundred yards to see the end. But the eagle could not quite make up its mind to attack. Twice it hovered within a foot or two of the yearling's head, again flew off and again returned. Finally the yearling trotted off after the rest of the band, and the eagle returned to the upper air. Later we found the carcass of a yearling, with two eagles, not to mention ravens and magpies, feeding on it; but I could not tell whether they had themselves killed the yearling or not.

Here and there in the region where the elk were abundant we came upon horses, which for some reason had been left out through the winter. They were much wilder than the elk. Evidently the Yellowstone Park is a natural nursery and breeding-ground of the elk, which here, as said above, far outnumber all the other game put together. In the winter, if they cannot get to open water, they eat snow; but in several places where there had been springs which kept open all winter, we could see by the tracks that they had been regularly used by bands of elk. The men working at the new road along the face of the cliffs beside the Yellowstone River near Tower Falls informed me that in October enormous droves of elk coming from the interior of the Park and travelling northward to the lower lands had crossed the Yellowstone just above Tower Falls. Judging by their description, the elk had crossed by thousands in an uninterrupted stream, the passage taking many hours. In fact nowadays these Yellowstone elk are, with the exception of the arctic caribou, the only American game which at times travel in immense droves like the buffalo of the old days.

A couple of days after leaving Cottonwood Creek—where we had spent several days—we camped at the Yellowstone Canyon below Tower Falls. Here we saw a second band of mountain-sheep, numbering only eight—none of them old rams. We were camped on the west side of the canyon; the sheep had their abode on the opposite side, where they had spent the winter. It has recently been customary among some authorities, especially the English hunters and naturalists who have written of the Asiatic sheep, to speak as if sheep were naturally creatures of the plains rather than mountain-climbers. I know nothing of the Old World sheep, but the Rocky Mountain bighorn is to the full as characteristic a mountain animal, in every sense of the word, as chamois, and, I think, as the ibex. These sheep were well known to the road-builders, who had spent the winter in the locality. They told me they never went back on the plains, but throughout the winter had spent their days and nights on the top of the cliff and along its face. This cliff was an alternation of sheer precipices and very steep inclines. When coated with ice it would be difficult to imagine an uglier bit of climbing; but throughout the winter, and even in the wildest storms, the sheep had habitually gone down it to drink at the water below. When we first saw them they were lying sunning themselves on the edge of the canyon, where the rolling grassy country behind it broke off into the sheer descent. It was mid-afternoon and they were under some pines. After a while they got up and began to graze, and soon hopped unconcernedly down the side of the cliff until they were half-way to the bottom. They then grazed along the sides, and spent some time licking at a place where there was evidently a mineral deposit. Before dark they all lay down again on a steeply inclined jutting spur midway between the top and bottom of the canyon.

Next morning I thought I would like to see them close up, so I walked down three or four miles below where the canyon ended, crossed the stream, and came up the other side until I got on what was literally the stamping-ground of the sheep. Their tracks showed that they had spent their time for many weeks, and probably for all the winter, within a very narrow radius. For perhaps a mile and a half, or two miles at the very outside, they had wandered to and fro on the summit of the canyon, making what was almost a well-beaten path; always very near and unusually on the edge of the cliff, and hardly ever going more than a few yards back into the grassy plain and hill country. Their tracks and dung covered the ground. They had also evidently descended into the depths of the canyon wherever there was the slightest break or even lowering in the upper line of the basalt cliffs. Although mountain-sheep often browse in winter, I saw but few traces of browsing here; probably on the sheer cliff-side they always get some grazing.

When I spied the band they were lying not far from the spot in which they had lain the day before, and in the same position on the brink of the canyon. They saw me and watched me with interest when I was two hundred yards off, but they let me get up within forty yards and sit down on a large stone to look at them, without running off. Most of them were lying down, but a couple were feeding steadily throughout the time I watched them. Suddenly one took the alarm and dashed straight over the cliff, the others all following at once. I ran after them to the edge in time to see the last yearling drop off the edge of the basalt cliff and stop short on the sheer slope below, while the stones dislodged by his hoofs rattled down the canyon. They all looked up at me with great interest, and then strolled off to the edge of a jutting spur and lay down almost directly underneath me and some fifty yards off. That evening on my return to camp we watched the band make its way right down the river-bed, going over places where it did not seem possible a four-footed creature could pass. They halted to graze here and there, and down the worst places they went very fast with great bounds. It was a marvelous exhibition of climbing.

After we had finished this horseback trip we went on sleds and skis to the upper Geyser Basin and the Falls of the Yellowstone. Although it was the third week in April, the snow was still several feet deep, and only thoroughly trained snow horses could have taken the sleighs along, while around the Yellowstone Falls it was possible to move only on snow-shoes. There was little life in those woods. In the upper basin I caught a meadow-mouse on the snow; I afterward sent it to Hart Merriam, who told me it was of a species he had described from Idaho, *Microtus nanus*; it had not been previously found in the Yellowstone region. We saw an occasional pine-squirrel, snow-shoe rabbit, or marten; and in the open meadows around the hot waters there were Canada geese and ducks of several species, and now and then a coyote. Around camp Clark's crows and Stellar's jays, and occasionally magpies, came to pick at the refuse; and of course they were accompanied by the whiskey-jacks, which behaved with their usual astounding familiarity. At Norris Geyser Basin there was a perfect chorus of bird music from robins, Western purple finches, juncos, and mountain-bluebirds. In the woods there were mountain-chickadees and pigmy nuthatches, together with an occasional woodpecker. In the northern country we had come across a very few blue grouse and ruffed grouse, both as tame as possible. We had seen a pigmy owl no larger than a robin sitting on the top of a pine in broad daylight, and uttering at short intervals a queer un-owl-like cry.

The birds that interested us most were the solitaires, and especially the dippers or water-ousels. We were fortunate enough to hear the solitaires sing not only when perched on trees but on the wing, soaring over a great canyon. They are striking birds in every way, and their habit of singing while soaring and their song are alike noteworthy. Once I heard a solitaire singing at the top of a canyon, and an ousel also singing but a thousand feet below him; and in this case I though the ousel sang better than his unconscious rival. The ousels are to my mind well-nigh the most attractive of all our birds, because of their song, their extraordinary habits, their whole personality. They stay through the winter in the Yellowstone because the waters are in many places open. We heard them singing cheerfully, their ringing melody having a certain suggestion of the winter wren's. Usually they sang while perched on some rock on the edge or in the middle of the stream; but sometimes on the wing; and often just before dipping under the torrent, or just after slipping out from it to some ledge of rock or ice. In the open places the Western meadow-lark was uttering its beautiful song; a real song as compared to the plaintive notes of its Eastern brother, and though short, yet with continuity and tune as well as melody. I love to hear the Eastern meadow-lark in the early spring; but I love still more the song of the Western meadow-lark. No bird escaped John Burroughs's eye; no bird note escaped his ear.

I cannot understand why the Old World ousel should have received such comparatively scant attention in the books, whether from nature writers or poets; whereas our ousel has greatly impressed all who know him. John Muir's description comes nearest doing him justice. To me he seems a more striking bird than, for instance, the skylark; though of course I not only admire but am very fond of the skylark. There are various pipits and larks in our own country which sing in highest air, as does the skylark, and their songs, though not as loud, are almost as sustained; and though they lack the finer kind of melody, so does his. The ousel, on the contrary, is a really brilliant singer, and in his habits he is even farther removed from the commonplace and the uninteresting than the lark himself. Some birds, such as the ousel, the mockingbird, the solitaire, show marked originality, marked distinction; others do not; the chipping-sparrow, for instance while in no way objectionable (like the imported house-sparrow), is yet a hopelessly commonplace little bird alike in looks, habits, and voice.

On the last day of my stay it was arranged that I should ride down from Mammoth Hot Springs to the town of Gardiner, just outside the Park limits, and there make an address at the laying of the corner-stone of the arch by which the main road is to enter the Park. Some three thousand people had gathered to attend the ceremonies. A little over a mile from Gardiner we came down out of the hills to the flat plain; from the hills we could see the crowd gathered around the arch, waiting for me to come. We put spurs to our horses and cantered rapidly toward the appointed place, and on the way we passed within forty yards of a score of blacktails, which merely moved to one side and looked at us, and within almost as short a distance of half a dozen antelope. To any lover of nature it could not help being a delightful thing to see the wild and timid creatures of the wilderness rendered so tame; and their tameness in the immediate neighborhood of Gardiner, on the very edge of the Park, spoke volumes for the patriotic good sense of the citizens of Montana. At times the antelope actually cross the Park line to Gardiner, which is just outside, and feed unmolested in the very streets of the town; a fact which shows how very far advanced the citizens of Gardiner are in right feeling on this subject; for of course the Federal laws cease to protect the antelope as soon as they are out of the Park. Major Pitcher informed me that both the Montana and Wyoming people were co-operating with him in zealous fashion to preserve the game and put a stop to poaching. For their attitude in this regard they deserve the cordial thanks of all Americans interested in these great popular playgrounds, where bits of the old wilderness scenery and the old wilderness life are to be kept unspoiled for the benefit of our children's children. Eastern people, and especially Eastern sportsmen, need to keep steadily in mind the fact that the Westerners who live in the neighborhood of the forest preserves are the men who in the last resort will determine whether or not these preserves are to be permanent. They cannot in the long run be kept as forest and game reservations unless the settlers round about believe in them and heartily support them; and the rights of these settlers must be carefully safe-guarded, and they must be shown that the movement is really in their interest. The Eastern sportsman who fails to recognize these facts can do little but harm by advocacy of the forest reserves.

It was in the interior of the Park, at the hotels beside the lake, the falls, and the various geyser basins that we would have seen the bears had the season been late enough; but unfortunately the bears were still for the most part hibernating. We saw two or three tracks, but the animals themselves had not yet begun to come about the hotels. Nor were the hotels open. No visitors had previously entered the Park in the winter or early spring, the scouts and other employees being the only ones who occasionally traverse it. I was sorry not to see the bears, for the effect of protection upon bear life in the Yellowstone has been one of the phenomena of natural history. Not only have they grown to realize that they are safe, but, being natural scavengers and foul feeders, they have come to recognize the garbage heaps of the hotels as their special sources of food-supply. Throughout the summer months they come to all the hotels in numbers, usually appearing in the late afternoon or evening, and they have become as indifferent to the presence of men as the deer themselves—some of them very much more indifferent. They have now taken their place among the recognized sights of the Park, and the tourists are nearly as much interested in them as in the geysers. In mussing over the garbage heaps they sometimes get tin cans stuck on their paws, and the result is painful. Buffalo Jones and some of the other scouts in extreme cases rope the bear, tie him up, cut the tin can of his paw, and let him go again. It is not an easy feat, but the astonishing thing is that it should be performed at all.

It was amusing to read the proclamations addressed to the tourists by the Park management, in which they were solemnly warned that the bears were really wild animals, and that they must on no account be either fed or teased. It is curious to think that the descendants of the great grizzlies which were the dread of the early explorers and hunters should now be semidomesticated creatures, boldly hanging around crowded hotels for the sake of what they can pick up, and quite harmless so long as any reasonable precaution is exercised. They are much safer, for instance, than any ordinary bull or stallion, or even ram, and, in fact, there is no danger from them at all unless they are encouraged to grow too familiar or are in some way molested. Of course among the thousands of tourists there is a percentage of fools; and when fools go out in the afternoon to look at the bears feeding they occasionally bring themselves into jeopardy by some senseless act. The black bears and the cubs of the bigger bears can readily be driven up trees, and some of the tourists occasionally do this. Most of the animals never think of resenting it; but now and then one is run across which has its feeling ruffled by the performance. In the summer of 1902 the result proved disastrous to a too inquisitive tourist. He was travelling with his wife, and at one of the hotels they went out toward the garbage pile to see the bears feeding. The only bear in sight was a large she, which, as it turned out, was in a bad temper because another party of tourists a few minutes before had been chasing her cubs up a tree. The man left his wife and walked toward the bear to see how close he could get. When he was some distance off she charged him, whereupon he bolted back toward his wife. The bear overtook him, knocked him down, and bit him severely. But the man's wife, without hesitation, attacked the bear with that thoroughly feminine weapon, an umbrella, and frightened her off. The man spent several weeks in the Park hospital before he recovered. Perhaps the following telegram sent by the manager of the Lake Hotel to Major Pitcher illustrates with sufficient clearness the mutual relations of the bears, the tourists, and the guardians of the public weal in the Park. The original was sent me by Major Pitcher. It runs:

> Lake. 7–27–'03. Major Pitcher, Yellowstone: As many as seventeen bears in an evening appear on my garbage dump. To-night eight or ten. Campers and people not of my hotel throw things at them to make them run away. I cannot, unless there personally, control this. Do you think you could detail a trooper to be there every evening from say six o'clock until dark and make people remain behind danger line laid out by Warden Jones? Otherwise I fear some accident. The arrest of one or two of these campers might help. My own guests do pretty well as they are told. James Barton Key. 9 A. M.

Major Pitcher issued the order as requested.

At times the bears get so bold that they take to making inroads on the kitchen. One completely terrorized a Chinese cook. It would drive him off and then feast upon whatever was left behind. When a bear begins to act in this way or to show surliness it is sometimes necessary to shoot it. Other bears are tamed until they will feed out of the hand, and will come at once if called. Not only have some of the soldiers and scouts tamed bears in this fashion, but occasionally a chambermaid or waiter girl at one of the hotels has thus developed a bear as a pet.

This whole episode of bear life in the Yellowstone is so extraordinary that it will be well worth while for any man who has the right powers and enough time to make a complete study of the life and history of the Yellowstone bears. Indeed, nothing better could be done by some of our outdoor faunal naturalists than to spend at least a year in the Yellowstone, and to study the life habits of all the wild creatures therein. A man able to do this, and to write down accurately and interestingly what he has seen, would make a contribution of permanent value to our nature literature.

# 9

# Bird Reserves at the Mouth of the Mississippi

ON JUNE 7, 1915, I WAS THE GUEST OF MY FRIEND JOHN M. Parker, of New Orleans, at his house at Pass Christian, Mississippi. For many miles west, and especially east, of Pass Christian, there are small towns where the low, comfortable, singularly picturesque and attractive houses are owned, some by Mississippi planters, some by city folk who come hither from the great Southern cities, and more and more in winter-time from the great Northern cities also, to pass a few months. The houses, those that are isolated and those in the little towns, stand in what is really one long row; a row broken by vacant reaches, but as a whole stretching for sixty miles, with the bright waters of the Gulf lapping the beach in front of them, and behind them leagues of pine forest. Between the Gulf and the waters lies a low ridge or beach of white sand. It is hard to make anything grow in this sand; but the owners of the houses have succeeded, using dead leaves and what manure is available; and in this leaf-mould the trees and grasses and flowers grow in profusion. Long, flimsy wooden docks stretch out into the waters of the Gulf; there is not much bad weather, as a rule, but every few years there comes a terrible storm which wrecks buildings and bridges, destroys human lives by the thousand, washes the small Gulf sailing craft ashore, and sweeps away all the docks.

Our host's house was cool and airy, with broad, covered verandas, and mosquito screens on the doors and the big windows. The trees in front were live-oaks, and others of his own planting—

magnolias, pecans, palms, and a beautiful mimosa. The blooming oleanders and hydrangeas were a delight to the eye. Behind, the place stretched like a long ribbon to the edge of the fragrant pine forest, where the long-leaved and loblolly pines rose like tall columns out of the needle-covered sand. Five pairs of mocking-birds and one pair of thrashers had just finished nesting; at dawn, when the crescent of the dying moon had risen above the growing light in the east, the mockers sang wonderfully, and after a while the thrasher chimed in. Only the singing of nightingales where they are plentiful, as in some Italian woods, can compare in strength and ecstasy and passion, in volume and intricate change and continuity, with the challenging love-songs of many mockers, rivalling one another, as they perch and balance and spring upward and float downward through the branches of live-oak or magnolia, after sunset and before sunrise, in the warm, still, brilliant moonlight of spring and early summer.

There were other birds. The soldierly looking red-headed woodpeckers, in their striking black, red, and white uniform, were much in evidence. Gaudy painted finches, or "non-pareils," were less conspicuous only because of their small size. Blue jays had raised their young in front of the house, and, as I was informed, had been successfully beaten off by the mockers and thrashers when they attempted assaults on the eggs and nestlings of the latter. Purple martins darted through the air. King-birds chased the big grackles and the numerous small fish-crows—not so very much bigger than the grackles—which uttered queer, hoarse croakings. A pair of crested flycatchers had their nest in a hollow in a tree; the five boldly marked eggs rested, as usual, partly on a shed snake skin. How, I wonder, through the immemorial ages, and why, did this particular bird develop its strange determination always, where possible, to use a snake's cast-off skin in building its nest? Every season, I was told this flycatcher nested in the same hollow; and every season the hollow was previously nested in by a tufted titmouse. Loggerhead shrikes were plentiful. Insects were their usual food, but they also pounced on small birds, mice, and lizards, and once on a little chicken. They empale their prey on locust thorns and on the spines of other trees and bushes; and I have known a barbed-wire fence to be decorated with the remains of their victims. There were red cardinal-birds; and we saw another red bird also, a summer tanager.

But the most interesting birds on the place were not wild, being nothing more nor less than ordinary fowls engaged in what to me were most unordinary occupations. Parker had several hundred fowls, and had by trial discovered the truth of the statement that capons make far better mothers than do hens, especially for very young chicks. We saw dozens of broods of chickens, and one or two of young guinea-fowl, being taken care of by caponized bantams, game-cocks, and cochin-chinas. These improvised mothers looked almost precisely as they did before being caponized, the differences, chiefly in the color of the comb, being insignificant, for they were full-grown birds when operated on. But their natures had suffered the most extraordinary change, for they had developed not only the habits but the voices of unusually exemplary mother hens. They never crowed; they clucked precisely like hens; and they protected, covered, fed, and led about their broods just like hens. They were timid, except in defense of the chicks; but on their behalf they were really formidable fighters. The change in habits takes place with extraordinary rapidity. In a few hours the cock has completely changed and can be placed with a brood which he promptly adopts. In perhaps one case in ten he does not take readily to his duties as an *ex-officio* hen; and in such case the further measure adopted seems as incredible as the rest of the performance, for he is made drunk with whiskey, acts as if he were intoxicated, and then promptly develops maternal feelings, and zealously enters on his new career.

We saw game-cocks clucking and calling to their broods of little chicks, to get them to the crumbs we tossed to them, and then sitting with the chicks not only under their wings but on their backs. They kept the broods with them until the young were nearly as large as they were; in one case the brood consisted of guinea-fowl. Moreover, they welcomed any brood, no matter how large. One big rooster was leading around so many chickens—all, by what seemed a sardonic jest, his own progeny, the progeny of the days when he was a mere unregenerate father—that when they took shelter under him he had to spread his wings; "like a buzzard," said my host, to whom soaring buzzards were familiar sights. Of course, the extraordinary part of all this was not the loss of the male qualities but the immediate and complete acquirement of those of the female. It was as if steers invariably took to mothering calves, or geldings to adopting foals.

These capon-mothers, with their weight and long spurs, fought formidably for their chicks. In one case a Cooper's hawk swooped on a half-grown chick, whereupon the game-cock who was officiating as hen flew at the aggressor, striking it so hard as to injure the top of the wing. The hawk was unable to fly, and the cock pressed it too close to let it escape. Although the rooster could not kill the hawk, for the latter threw itself on its back with extended talons, he had rendered it unable to escape, and one of the men about the place came up and killed it, having been attracted by the noise of the fight. Another cock killed a big blacksnake which tried to carry off one of the chicks. The cock darted to and fro over the snake, striking it continually until it succumbed.

Pass Christian is an ideal place for a man to go who wishes to get away from the Northern cold for a few weeks, and be where climate, people, and surroundings are all delightful, and the fishing and shooting excellent. There is a good chance, too, that the fish and game will be preserved for use, instead of recklessly exterminated; for during the last dozen years Louisiana and Mississippi, like the rest of the Union, have waked to the criminality of marring and ruining a beautiful heritage which should be left and through wise use (not nonuse) can be left, undiminished, to the generations that are to come after us. As yet the Gulf in front of the houses swarms with fish of many kinds up to the great tarpon, the mailed and leaping giant of the warm seas; and with the rapid growth of wisdom in dealing with nature we may hope that there will soon be action looking toward the regulation of seining and to protection of the fish at certain seasons. On land the quail have increased in the neighborhood of Pass Christian during the last few years. This is largely due to the activity of my host and his two sons as hunters. They have a pack of beagles, trained to night work, and this pack has to its credit nearly four hundred coons and possums—together with an occasional skunk!—and, moreover, has chivied the gray foxes almost out of the country; and all these animals are the inveterate enemies of all small game, and especially of ground-nesting birds. To save interesting creatures, it is often necessary not merely to refrain from killing them but also to war on their enemies.

One of the sons runs the Parker stock-farm in upper Louisiana, beside the Mississippi. There are about four thousand acres, half of it highland, the other half subject to flood if the levees break. Five years ago such a break absolutely destroyed the Parker plantations, then exclusively on low land. Now, in event of flood, the stock can be driven, and the human beings escape, to the higher ground. Young Parker, now twenty-two years old, has run the plantation since he was sixteen. The horses, cattle, and sheep are all of the highest grade; the improvement in the stock of Louisiana and Mississippi during the last two decades has been really noteworthy. Game, and wild things generally, have increased in numbers on this big stock-farm. There is no wanton molestation of any animal permitted, no plundering of nests, no shooting save within strictly defined limits, and so far as possible all rare things are given every chance to increase. As an example, when, in clearing a tract of swamp land, a heron's nest was discovered, the bushes round about were left undisturbed, and the heron family was reared in safety. Wild turkeys have somewhat, and quail very markedly, increased. The great horned owls, which destroyed the ducks, have to be warred against, and the beasts of prey likewise. Surely it will ultimately again be recognized in our country that life on a plantation, on a great stock-farm or ranch, is one of the most interesting, and, from the standpoint of both body and soul, one of the most healthy, of all ways of earning a living.

At four on the morning of the 8th our party started from the wharf in front of Pass Christian. We were in two boats. One, good-sized and comfortable, under the command of Captain Lewis Young, was the property of the State Conservation Commission of Louisiana, the commission having most courteously placed it at our disposal. On this boat were my host, his two sons, John, Jr., and Tom, myself, and a photographer, Mr. Coquille, of New Orleans. The other boat, named the *Royal Tern*, was the property of the Audubon Society, being allotted to the work of cruising among and protecting the bird colonies on those islands set apart as bird refuges by the National and State Governments. On this boat—which had a wretched engine, almost worthless—went Mr. Herbert K. Job and Mr. Frank M. Miller. Mr. Miller was at one time president of the Louisiana Conservation Commission, and the founder of the Louisiana State Audubon Society, and is one of the group of men to whom she owes it that she, the home state of Audubon, of our first great naturalist, is now thoroughly awake to the danger of reckless waste and destruction of all the natural resources of the State, including the birds. Mr. Herbert K. Job is known to all who care for bird study and bird preservation. He is a naturalist who has made of bird-photography a sport, a science, and an art. His pictures, and his books in which these pictures appear, are fascinating both to the scientific ornithologist and to all lovers of the wild creatures of the open. Like the other field naturalists I have known, like the men who were with me in Africa and South America, Mr. Job is an exceptionally hardy, resolute, and resourceful man, following his wilderness work with single-minded devotion, and continually, and in matter-of-fact manner, facing and overcoming hardship, wearing toil, and risk which worthy stay-at-home people have no means whatever of even gauging. I owed the pleasure of Mr. Job's company to Mr. Frank M. Chapman, at whose suggestion he was sent with me by the National Audubon Society.

The State Conservation Commission owes its existence to the wise public spirit and farsightedness of the Louisiana Legislature. The Audubon Society, which has done far more than any other single agency in creating and fostering an enlightened public sentiment for the preservation of our useful and attractive birds, is a purely voluntary organization, consisting of men and women who in these matters look further ahead than their fellows, and who have the precious gift of sympathetic imagination, so that they are able to see, and to wish to preserve for their children's children, the beauty and wonder of nature. (During the year preceding this trip, by the way, the society enrolled one hundred and fifty-one thousand boys and girls in its junior bird clubs, all of which give systematic instruction in the value of bird life.) It was the Audubon Society which started the movement for the establishment of bird refuges. The society now protects and polices about one hundred of these refuges, which, of course, are worthless unless thus protected.

The *Royal Tern* is commanded by Captain William Sprinkle, born and bred on this Gulf coast, who knows the sea-fowl, and the islands where they breed and dwell, as he knows the winds and the lovely, smiling, treacherous Gulf waters. He is game warden , and he and the *Royal Tern* are the police force for over five hundred square miles of sand-bars, shallow waters, and intricate channels. The man and the boat are two of the chief obstacles in the way of the poachers, the plume-hunters, and eggers, who always threaten these bird sanctuaries.

Many of these poachers are at heart good men, who follow their fathers' business, just as respectable men on the seacoast once followed the business of wrecking. But when times change and a once acknowledged trade comes under the ban of the law the character of those following it also changes for the worse. Wreckers are no longer respectable, and plume-hunters and eggers are sinking to the same level. The illegal business of killing breeding birds, of leaving nestlings to starve wholesale, and of general ruthless extermination, more and more tends to attract men of the same moral category as those who sell whiskey to Indians and combine the running of "blind pigs" with highway robbery and murder for hire.

In Florida one of the best game wardens of the Audubon Society was killed by these sordid bird-butchers. A fearless man and a good boat are needed to keep such gentry in awe. Captain Sprinkle meets the first requirement, the hull of the *Royal Tern* the second. But the engines of the *Tern* are worthless; she can catch no freebooter; she is safe only in the mildest weather. Is there not some bird-lover of means and imagination who will put a good engine in her? Such a service would be very real. As for Caption Sprinkle, his services are, of course, underpaid, his salary bearing no relation to their value. The Biological Survey does its best with its limited means; the Audubon Society adds something extra; but this very efficient and disinterested laborer is worth a good deal more than the hire he receives. The government pays many of its servants, usually those with rather easy jobs, too much; but the best men who do the hardest work, the men in the life-saving and lighthouse service, the forest-rangers, and those who patrol and protect the reserves of wild life, are almost always underpaid.

Yet, in spite of all the disadvantages, much has been accomplished. This particular reservation was set apart by presidential proclamation in 1905. Captain Sprinkle was at once put in charge. Of the five chief birds, the royal terns, Caspian terns, Cabot's terns, laughing gulls, and skimmers, there were that season about one thousand nests. This season, ten years later, there are about thirty-five thousand nests. The brown pelicans and Louisiana herons also show a marked increase. The least tern, which had been completely exterminated or driven away, has returned and is breeding in fair numbers.

As we steamed away from the Pass Christian dock dawn was turning to daylight under the still brilliant crescent moon. Soon we saw the red disk of the sun rising behind the pine forest. We left Mississippi Sound, and then were on the Gulf itself. The Gulf was calm, and the still water teemed with life. Each school of mullets or sardines could be told by the queer effect on the water, as of a cloud shadow. Continually we caught glimpses of other fish; and always they were fleeing from death or ravenously seeking to inflict death on the weak. Nature is ruthless, and where her sway is uncontested there is no peace save the peace of death; and the fecund stream of life especially of life on the lower levels, flows like an immense torrent out of non-existence for but the briefest moment before the enormous majority of the beings composing it are engulfed in the jaws of death, and again go out into the shadow.

Huge rays sprang out of the water and fell back with a resounding splash. Devil-fish, which made the rays look like dwarfs, swam slowly near the surface; some had their mouths wide open as they followed their prey. Globular jellyfish, as big as pumpkins, with translucent bodies, pulsed through the waters; little fishes and crabs swam among their short, thick tentacles and in between the waving walls into which the body was divided. Once we saw the head of a turtle above water; it was a logger-head turtle, and the head was as large as the head of a man; when I first saw it, above the still water, I had no idea what it was.

By noon we were among the islands of the reservation. We had already passed other and larger islands, for the most part well wooded. On these there were great numbers of coons and minks, and therefore none of the sea-birds which rest on the ground or in low bushes. The coons are more common than the minks and muskrats. In the inundations they are continually being carried out to sea on logs; a planter informed me that on one occasion in a flood he met a log sailing down the swollen Mississippi with no less than eleven coons aboard. Sooner or later castaway coons land on every considerable island off the coast, and if there is fresh water, and even sometimes if there is none, they thrive; and where there are many coons, the gulls, terns, skimmers, and other such birds have very little chance to bring up their young. Coons are fond of rambling along beaches; at low tide they devour shell-fish; and they explore the grass tufts and bushes, and eat nestlings, eggs, and even the sitting birds. If on any island we found numerous coon tracks there were usually few nesting sea-fowl, save possibly on some isolated point. The birds breed most plentifully in the numberless smaller islands—some of considerable size—where there is no water, and usually not a tree. Some of these islands are nothing but sand, with banks and ramparts of shells, while others are fringed with marsh-grass and covered with scrub mangrove. But the occasional fierce tropical storms not only change the channels and alter the shape of many of the islands, but may even break up some very big island. In such case an island with trees and water may for years be entirely uninhabited by coons, and the birds may form huge rookeries thereon. The government should exterminate the coons and minks on all the large islands, so as to enable the birds to breed on them; for on the small islands the storms and tides work huge havoc with the nests.

Captain Young proved himself not only a first-class captain but a first-class pilot through the shifting and tangled maze of channels and islands. The *Royal Tern*, her engines breaking down intermittently, fell so far in the rear that in the early afternoon we anchored, to wait for her, off an island to which a band of pelicans resorted—they had nested, earlier in the year, on another island some leagues distant. The big birds, forty or thereabouts in number, were sitting on a sand-spit which projected into the water, enjoying a noontide rest. As we approached they rose and flapped lazily out to sea for a few hundred yards before again lighting. Later in the afternoon they began to fly to the fishing-grounds, and back and forth, singly and in small groups. In flying they usually gave a dozen rapid wing-beats, and then sailed for a few seconds. If several were together the leader gave the "time" to the others; they all flapped together, and then all glided together. The neck was carried in a curve, like a heron's; it was only stretched out straight like a stork's or bustard's when the bird was diving. Some of the fishing was done, singly or in parties, in the water, the pelicans surrounding shoals of sardines and shrimps, and scooping them up in their capacious bags. But, although such a large, heavy bird, the brown pelican is an expert wing-fisherman also. A pair would soar round in circles, the bill perhaps pointing downward, instead of, as usual, being held horizontally. Then, when the fish was spied the bird plunged down, almost perpendicularly, the neck stretched straight and rigid, and disappeared below the surface of the water with a thump and splash, and in a couple of seconds emerged, rose with some labor, and flew off with its prey. At this point the pelicans had finished breeding before my arrival—although a fortnight later Mr. Job found thousands of fresh eggs in their great rookeries west of the mouth of the Mississippi. The herons had well-grown nestlings, whereas the terns and gulls were in the midst of the breeding, and the skimmers had only just begun. The pelicans often flew only a few yards, or even feet, above the water, but also at times soared or wheeled twenty or thirty rods in the air, or higher. They are handsome, interesting birds and add immensely, by their presence, to the pleasure of being out on these waters; they should be completely protected everywhere—as, indeed, should most of these sea-birds.

The two Parker boys—the elder of whom had for years been doing a man's work in the best fashion, and the younger of whom had just received an appointment to Annapolis—kept us supplied with fish, caught with the hook and rod, except the flounders, which were harpooned. The two boys were untiring; nothing impaired their energy, and no chance of fatigue and exertion, at any time of the day or night, appealed to them save as an exhilarating piece of good fortune. At a time when so large a section of our people, including especially those who claim in a special sense to be the guardians of cultivation, philanthropy, and religion, deliberately make a cult of pacifism, poltroonery, sentimentality, and neurotic emotionalism, it was refreshing to see the fine, healthy, manly young fellows who were emphatically neither "too proud to fight" nor too proud to work, and with whom hard work, and gentle regard for the rights of others, and the joy of life, all went hand in hand.

Toward evening of our first day the weather changed for the worse; the fishers among the party were recalled, and just before nightfall we ran off, and after much groping in the dark we made a reasonably safe anchorage. But midnight the wind fell, dense swarms of mosquitoes came aboard, and, as our mosquito-nets were not well up (thanks partly to our own improvidence, and partly to the violence of the wind, for we were sleeping on deck because of the great heat), we lived in torment until morning. On the subsequent nights we fixed our mosquito-bars so carefully that there was no trouble. Mosquitoes and huge, green-headed horse-flies swarm on most of the islands. I witnessed one curious incident in connection with one of these big, biting horse-flies. A kind of wasp preys on them, and is locally known as the "horse-guard," or "sheriff-fly," accordingly. These horse-guards are formidable-looking things and at first rather alarm strangers, hovering round them and their horses; but they never assail beast or man unless themselves molested, when they are ready enough to use their powerful sting. The horses and cattle speedily recognize these big, humming, hornet-like horse-guards as the foes of their tormentors. As we walked over the islands, and the green-headed flies followed us, horse-guards also joined us; and many greenheads and some horse-guards came on board. Usually when the horse-guard secured the greenhead it was pounced on from behind, and there was practically no struggle—the absence of struggle being usual in the world of invertebrates, where the automaton-like actions of both preyer and prey tend to make each case resemble all others in its details. But on one occasion the greenhead managed to turn, so that he fronted his assailant and promptly grappled with him, sinking his evil lancet into the wasp's body and holding the wasp so tight that the latter could not thrust with its sting. They grappled thus for several minutes. The horse-guard at last succeeded in stabbing its antagonist, and promptly dropped the dead body. Evidently it had suffered much, for it vigorously rubbed the wounded spot with its third pair of legs, walked hunched up, and was altogether a very sick creature.

On the following day we visited two or three islands which the man-of-war birds were using as roosts. These birds are the most wonderful fliers in the world. No other bird has such an expanse of wing in proportion to the body weight. No other bird of its size seems so absolutely at home in the air. Frigate-birds—as they are also called—hardly ever light on the water, yet they are sometimes seen in mid-ocean. But they like to live in companies near some coast. They have very long tails, usually carried closed, looking like a marlin-spike, but at times open, like a great pair of scissors, in the course of their indescribably graceful aerial evolutions. We saw them soaring for hours at a time, sometimes to all seeming absolutely motionless as they faced the wind. They sometimes caught fish for themselves, just rippling the water to seize surface swimmers, or pouncing with startling speed on any fish which for a moment leaped into the air to avoid another shape of ravenous death below. If the frigate-bird caught the fish transversely, it rose, dropped its prey, and seized it again by the head before it struck the water. But it also obtained its food in less honorable fashion—by robbing other birds. The pelicans were plundered by *all* their fish-eating neighbors, even the big terns; but the man-of-war bird robbed the robbers. We saw three chase a royal tern, a very strong flier; the tern towered, ascending so high we could hardly see it, but in great spirals its pursuers rose still faster, until one was above it; and then the tern dropped the fish, which was snatched in mid-air by one of the bandits. Captain Sprinkle had found these frigate-birds breeding on one of the islands the previous year, each nest being placed in a bush and containing two eggs. We visited the island; the big birds—the old males jet black, the females with white breasts, the young males with white heads—were there in numbers, perched on the bushes, and rising at our approach. But there were no nests, and, although we found one fresh egg, it was evidently a case of sporadic laying, having nothing to do with home-building.

On another island, where we also found a big colony of frigate-birds roosting on the mangrove and Gulf tamarisk scrub, there was a small heronry of the Louisiana heron. The characteristic flimsy heron nests were placed in the thick brush, which was rather taller than a man's head. The young ones had left the nests, but were still too young for anything in the nature of sustained flight. They were, like all young herons, the pictures of forlorn and unlovely inefficiency, as they flapped a few feet away and strove with ungainly awkwardness to balance themselves on the yielding bush tops. The small birds we found on the islands were red-winged blackbirds, Louisiana seaside sparrows, and long-billed marsh-wrens—which last had built their domed houses among the bushes, in default of tall reeds. On one island Job discovered a night-hawk on her nest. She fluttered off, doing the wounded-bird trick, leaving behind her an egg and a newly hatched chick. He went off to get his umbrella-house, and when he returned the other egg was hatching, and another little chick, much distressed by the heat, appeared. He stood up a clamshell to give it shade, and then, after patient waiting, the mother returned, and he secured motion-pictures of her and her little family. These birds offer very striking examples of real protective coloration.

The warm shallows, of course, teem with mollusks as well as with fish—not to mention the shrimps, which go in immense silver schools, and which we found delicious eating. The occasional violent storms, when they do not destroy islands, throw up on them huge dikes or ramparts of shells, which makes the walking hard on the feet.

There are more formidable things than shells in the warm shallows. The fishermen as they waded near shore had to be careful lest they should step on a sting-ray. When a swim was proposed as our boat swung at anchor in mid-channel, under the burning midday sun, Captain Sprinkle warned us against it because he had just seen a large shark. He said that sharks rarely attacked men, but that he had known of two instances of their doing so in Mississippi Sound, one ending fatally. In this case the man was loading a sand schooner. He was standing on a scaffolding, the water half-way up his thighs, and the shark seized him and carried him into deep water. Boats went to his assistance at once, scaring off the shark; but the man's leg had been bitten nearly in two; he sank, and was dead when he was finally found.

The following two days we continued our cruise. We steamed across vast reaches of open Gulf, the water changing from blue to yellow as it shoaled. Now and then we sighted or passed low islands of bare sand and scrub. The sky was sapphire, the sun splendid and pitiless, the heat sweltering. We came across only too plain evidence of the disasters always hanging over the wilderness folk. A fortnight previously a high tide and a heavy blow had occurred coincidentally. On the islands where the royal terns especially loved to nest the high water spelled destruction. The terns nest close together, in bird cities, so to speak, and generally rather low on the beaches. On island after island the waves had washed over the nests and destroyed them by the ten thousand. The beautiful royal terns were the chief sufferers. On one island there was a space perhaps nearly an acre in extent where the ground was covered with their eggs, which had been washed thither by the tide; most of them had then been eaten by those smart-looking highwaymen, the trim, slate-headed laughing gulls. The terns had completely deserted the island and had gone in their thousands to another; but some skimmers remained and were nesting. The westernmost island we visited was outside the national reservation, and that very morning it had been visited and plundered by a party of eggers. The eggs had been completely cleared from most of the island, gulls and terns had been shot, and the survivors were in a frantic state of excitement. It was a good object-lesson in the need of having reserves, and laws protecting wild life, and a sufficient number of efficient officers to enforce the laws and protect the reserves. Defenders of the short-sighted men who in their greed and selfishness will, if permitted, rob our country of half its charm by their reckless extermination of all useful and beautiful wild things sometimes seek to champion them by saying that "the game belongs to the people." So it does; and not merely to the people now alive, but to the unborn people. The "greatest good of the greatest number" applies to the number within the womb of time, compared to which those now alive form but an insignificant fraction. Our duty to the whole, including the unborn enerations, bids us restrain an unprincipled present-day minority m wasting the heritage of these unborn generations. The movet for the conservation of wild life, and the larger movement conservation of all our natural resources, are essentially atic in spirit, purpose, and method.

ne of the islands we found where green turtles had crawled

up the beaches to bury their eggs in the sand. We came across two such nests. One of them I dug up myself. The eggs we took to the boat, where they were used in making delicious pancakes, which went well with fresh shrimp, flounder, weakfish, mackerel, and mullet.

The laughing gulls and the black skimmers were often found with their nests intermingled, and they hovered over our heads with the same noisy protest against our presence. Although they often—not always—nested so close together, the nests were in no way alike. The gulls' dark-green eggs, heavily blotched with brown, two or three in number, lay on a rude platform of marsh-grass, which was usually partially sheltered by some bush or tuft of reeds, or, if on wet ground, was on a low pile of driftwood. The skimmers' eggs, light whitish green and less heavily marked with brown, were, when the clutch was full, four to six in number. There was no nest at all, nothing but a slight hollow in the sand, or gravel or shell débris. In the gravel or among the shell débris it was at first hard to pick out the eggs; but as our eyes grew accustomed to them we found them without difficulty. Sometimes we found the nests of gull and skimmer within a couple of feet of one another, one often under or in a bush, the other always out on the absolutely bare open. Considering the fact that the gull stood ready, with cannibal cheerfulness, to eat the skimmer's eggs if opportunity offered, I should have thought that to the latter bird such association would have seemed rather grewsome; but, as a matter of fact, there seemed to be no feeling of constraint whatever on either side, and the only fighting I saw, and this of a very mild type, was among the gulls themselves. As we approached their nesting-places all these birds rose, and clamored loudly as they hovered over us, lighting not far off, and returning to their nests as we moved away.

The skimmers are odd, interesting birds, and on the whole were, if anything, rather tamer even than the royal terns and laughing gulls, their constant associates. They came close behind these two in point of abundance. They flew round and round us, and to and fro, continually uttering their loud single note, the bill being held half open as they did so. The lower mandible, so much longer than the upper, gives them a curious look. Ordinarily the bill is held horizontally and closed; but when after the small fish on which they feed the lower mandible is dropped to an angle of forty-five degrees, ploughing lightly the surface of the water and scooping up the prey. They fly easily, with at ordinary times rather deliberate strokes of their long wings, wheeling and circling, and continually crying if roused from their nests. When flying the white of their plumage is very conspicuous, and as they flapped around every detail of form and coloration, of bill and plumage, could be observed.

When sitting they appear almost black, and, in consequence, when on their nests, on the beaches or on the white-shell dikes, they are visible half a mile off, and stand out as distinctly as a crow on a snow-bank.[1] They are perfectly aware of this, and make no attempt to elude observation, any more than the gulls and terns do. The fledglings are concealingly colored and crouch motionless, so as to escape notice from possible enemies; and the eggs, while they do not in color harmonize with the surroundings to the extent that they might artificially be made to do, yet easily escape the eye when laid on a beach composed of broken sea-shells. But the coloration of the adults is of a strikingly advertising character, under all circumstances, and especially when they are sitting on their nests. Among all the vagaries of the fetichistic school of concealing-colorationists none is more amusing than the belief that the coloration of the adult skimmer is ever, under any conditions, of a concealing quality. Sometimes the brooding skimmer attempted to draw us away from the nest by fluttering off across the sand like a wounded bird. Like the gulls, the skimmers moved about much more freely on the ground than did the terns.

The handsome little laughing gull was found everywhere, and often in numerous colonies although these colonies were not larger than those of the skimmer, and in no way approached the great breeding assemblages of the royal terns on the two or three islands where the latter especially congregated. They were noisy birds, continually uttering a single loud note, but only occasionally the queer laughter which gives them their name. They looked very trim and handsome, both on the wing and when swimming or walking; and their white breasts and dark heads made them very conspicuous on their nests, no matter whether these were on open ground or partially concealed in a bush or reed cluster. Like the skimmers, although perhaps not quite so markedly, their coloration was strongly advertising at all times, including when on their nests. Their relations with their two constant associates and victims, the skimmer and the royal tern—the three being about the same size—seemed to me very curious. The gull never molested the eggs if either of the other birds if the parents were sitting on them or were close by. But gulls continually broke and devoured eggs, especially terns' eggs, which had been temporarily abandoned. Nor was this all. When a colony of nesting royal terns flew off at our approach, the hesitating advent of the returning parents was always accompanied by the presence of a few gulls. Commonly the birds lit a few yards away from the eggs, on the opposite side from the observer, and then by degrees moved forward among the temporarily forsaken eggs. The gulls were usually among the foremost ranks, and each, as it walked or ran to and fro, would now and then break or carry off an egg; yet I never saw a tern interfere or seem either alarmed or angered. These big terns are swifter and better fliers than the gulls, and the depredations take place all the time before their eyes. Yet they pay no attention that I could discern to the depredation. Compare this with the conduct of king-birds to those other egg-robbers, the crows. Imagine a king-bird, or, for that matter, a mocking-bird or thrasher, submitting with weak good humor to such treatment! If these big terns had even a fraction of the intelligence and spirit of king-birds, no gull would venture within a half-mile of their nesting-grounds.

It is one of the innumerable puzzles of biology that the number of eggs a bird lays seems to have such small influence on the abundance of the species. A royal tern lays one egg, rarely two; a gull three; a skimmer four to six. The gull eats the eggs of the other two, especially of the tern; as far as we know, all have the same foes; yet the abundance of the birds is in inverse ratio to the number of their eggs. Of course, there is an explanation; but we cannot even guess at it as yet. With this, as with so many other scientific questions, all we can say is, with Huxley, that we are not afraid to announce that we do not know.

The beautiful royal terns were common enough, flying in the air and diving boldly after little fish. We listened with interest to their cry, which was a kind of creaking bleat. We admired the silver of their plumage as they flew overhead. But we did not come across vast numbers of them assembled for breeding until the fourth day. Then we found them on an island on which Captain Sprinkle told us he had never before found them, although both skimmers and gulls had always nested on it. The previous fall he had waged war with the traps against the coons, which, although there was no fresh water, had begun to be plentiful on the island. He had caught a number, two escaping, one with the loss of a hind foot, and one with the loss of a fore foot. The island was seven miles long, curved, with occasional stretches of salt marsh, and with reaches of scrub, but no trees. Most of it was bare sand. We saw three coon tracks, two being those of the three-footed animals; evidently the damaged leg was now completely healed and was used like the others, punching a round hold in the sand. We saw one coon, at dusk, hunting for oysters at the water's edge.

The gulls and skimmers were nesting on this island in great numbers, but the terns were many times more plentiful. There were thousands upon thousands of them. Their breeding-places were strung in a nearly straight line for a couple of miles along the sand flats. A mile off, from our boat, we were attracted by their myriad forms, glittering in the brilliant sunlight as they rose and fell and crossed and circled over the nesting-places. The day was bright and hot, and the sight was one of real fascination. As we approached a breeding colony the birds would fly up, hover about, and resettle when we drew back a sufficient distance. The eggs, singly, or rarely in pairs, were placed on the bare sand, with no attempt at a nest, the brooding bird being sometimes but a few inches, sometimes two or three feet, from the nearest of its surrounding neighbors. The colonies of breeders were scattered along the shore for a couple of miles, each one being one or two hundred yards, or over, from the next. In one such breeding colony I counted a little over a thousand eggs; there were several of smaller size, and a few that were larger, one having perhaps three times as many. A number of the eggs, perhaps ten per cent, had been destroyed by the gulls; the coons had ravaged some of the gulls' nests, which were in or beside the scrub. The eggs of the terns, being so close together and on the bare sand, were very conspicuous; they were visible to a casual inspection at a distance of two or three hundred yards, and it was quite impossible for any bird or beast to overlook them near by. These gregarious nesters, whose eggs are gathered in a big nursery, cannot profit by any concealing coloration of the eggs. The eggs of the royal and Cabot's terns were perhaps a shade less conspicuous than the darker eggs of the Caspian tern, all of them lying together; but on that sand, and crowded into such a regular nursery, none of them could have escaped the vision of any foe with eyes. As I have said, the eggs of the skimmer, as the clutches were more scattered, were much more difficult to make out, on the shell beaches. Concealing coloration has been a survival factor only as regards a minority, and is responsible for the precise coloration of only a small minority, of adult birds and mammals; how much and what part it plays, and in what percentage of cases, in producing the coloration of eggs, is a subject which is well worth serious study. As regards most of these seabirds which nest gregariously, their one instinct for safety at nesting time seems to be to choose a lonely island. This is their only, and sufficient, method of outwitting their foes at the crucial period of the lives.

We found only eggs in the nurseries, not young birds. In each nursery there were always a number of terns brooding their eggs, and the air above was filled with a ceaseless flutter and flashing of birds leaving their nests and returning to them—or eggs, rather, for speaking accurately, there were no nests. The sky above was alive with the graceful, long-winged things. As we approached the nurseries the birds would begin to leave. If we halted before the alarm became universal, those that stayed always served as lures to bring back those that had left. If we came too near, the whole party rose in a tumult of flapping wings; and when all had thus left it was some time before any returned. With patience it was quite possible to get close to the sitting birds; I noticed that in the heat many had their bills open. Those that were on the wing flew round and round us, creaking and bleating, and often so near that every detail of form and color was vivid in our eyes. The immense majority were royal terns, big birds with orange beaks. With them were a very few Caspian terns, still bigger, and with bright-red beaks, and quite a number of Cabot's terns, smaller birds with yellow-tipped black beaks. These were all nesting together, in the same nurseries.

It has been said on excellent authority that terns can always be told from gulls because whereas the latter carry their beaks horizontally, the terns carry their bills pointing downward, "like a mosquito." My own observations do not agree with this statement. When hovering over water where there are fish, and while watching for their prey, terns point the bill downward, just as pelicans do in similar circumstances; just as gulls often do when they are seeking to spy food below them. But normally, on the great majority of the occasions when I saw them, the terns, like the gulls, carried the bill in the same plane as the body.

On another island we found a small colony of Forster's tern; and we saw sooty terns, and a few of the diminutive least terns. But I was much more surprised to find on, or rather over, one island a party of black terns. As these are inland birds, most of which at this season are breeding around the lakes of our Northwestern country, I was puzzled by their presence. Still more puzzling was it to come across the party of turnstones, with males in full, brightly varied nuptial dress, for the turnstones during the breeding season live north of the arctic circle, in the perpetual sunlight of the long polar day. On the other hand, a couple of big oyster-catchers seemed, and were, entirely in place; they are striking birds and attract attention at a great distance. We saw dainty Wilson's plover with their chicks, and also semipalmated sandpipers.

On the morning of the 12th we returned to Pass Christian. I was very glad to have seen this bird refuge. With care and protection the birds will increase and grow tamer and tamer, until it will be possible for any one to make trips among these reserves and refuges, and to see as much as we saw, at even closer quarters. No sight more beautiful and more interesting could be imagined.

I am far from disparaging the work of the collector who is also a field naturalist. On the contrary, I fully agree with Mr. Joseph Grinnell's recent plea for him. His work is indispensable. It is far more important to protect his rights than to protect those of the sportsman; for the serious work of the collector is necessary in order to prevent the scientific study of ornithology from lapsing into mere dilettanteism indulged in as a hobby by men and women with opera-glasses. Moreover, sportsmen also have their rights, and it is folly to sacrifice these rights to mere sentimentality—for, of course, sentimentality is as much the antithesis and bane of healthy sentiment as bathos is of pathos. If thoroughly protected, any bird or mammal would speedily increase in numbers to such a degree as to drive man from the planet; and of recent years this has been signally proved by actual experience as regard certain creatures, notably as regards the wapiti in the Yellowstone (where the prime need now is to provide for the annual killing of at least five thousand), and to a less extent as regards deer in Vermont.

But as yet these cases are rare exceptions. As yet with the great majority of our most interesting and important wild birds and beasts the prime need is to protect them, not only by laws limiting the open season and the size of the individual bag, but especially by the creation of sanctuaries and refuges. And, while the work of the collector is still necessary, the work of the trained faunal naturalist, who is primarily an observer of the life histories of the wild things, is even more necessary. The progress made in the United States, of recent years, in creating and policing bird refuges, has been of capital importance.

At nightfall of the third day of our trip, when we were within sight of Fort Jackson and of the brush and low trees which here grow alongside the Mississippi, we were joined by Mr. M. L. Alexander, the president of the Conservation Commission, on the commission's boat *Louisiana*. He was more than kind and courteous, as were all my Louisiana friends. He and Mr. Miller told me much of the work of the commission; work not only of the utmost use to Louisiana, but of almost equal consequence to the rest of the country, if only for the example set.

The commission was not founded until 1912, yet it has already accomplished a remarkable amount along many different lines. The work of reforestation of great stretches of denuded, and at present worthless, pine land has begun; work which will turn lumbering into a permanent Louisiana industry by making lumber a permanent crop asset, like corn or wheat, only taking longer to mature—an asset which it is equally important not to destroy. In taking care of the mineral resources a stop has been put to waste as foolish as it was criminal; for example, a gas-well which had flowed to waste until six million dollars' worth of gas had been lost was stopped and stored at the cost of five thousand one hundred dollars. The oysters are now farmed and husbanded, the beds being leased in such fashion that there is a steady improvement of the product. Louisiana is peculiarly rich in fish, and a policy has been inaugurated which, if preserved in, will make the paddle-fish industry as important as the sturgeon fishery is in Russia. Not only do the waters of Louisiana now belong to the State, but also the land under the water, this last proving in practise an admirable provision. Some three hundred thousand acres of game reserves and wild-life refuges (mostly uninhabitable by man) have now been established. These have largely been gifts to the State by wise and generous private individuals and corporations, the chief donors being Messrs. Edward A. McIlhenny and Charles Willis Ward, Mrs. Russell Sage, and the Rockefeller Foundation. The Conservation Commission has accepted the gifts, and is taking care of the reserves and refuges through its State wardens, with the result that wild birds of many kinds, including even the wary geese, which come down as winter visitants by the hundred thousand, have become very tame, and many beautiful birds which were on the verge of extinction are now re-established and increasing in numbers. These reserves, which lie for the most part in the low country along the coast, are west of the Mississippi.

Job had just come from a visit to the private reserve of Edward A. McIlhenny on Avery Island. It is the most noteworthy reserve in the country. It includes four thousand acres, and is near the Ward-McIlhenny reserve, which they have given to the State—a king's gift! Avery's Island is very beautiful. A great, shallow, artificial lake, surrounded by dwellings, fields, lawns, a railroad, and ox-wagon road, does not seem an ideal home for herons; but it has proved such under the care of Mr. McIlhenny. He started the reserve twenty years ago with eight snowy herons. Now it contains about forty thousand herons of several species. Complete freedom from molestation has rendered the birds extraordinarily tame. The beautiful snow-white lesser egret, which had been almost exterminated by the plume-hunters, flourishes by the thousand; the greater egret has been bothered so by the smaller one that it has retired before it; its heronries are now to be found mainly in other parts of the protected region. Many other kinds of heron, and many waterfowl, literally throng the place. Ducks winter by the thousand, and, most unexpectedly, some even of the northern kinds, like the gadwall, now stay to breed. Most of these birds are so tame that there is little difficulty in taking photographs of them.

The Audubon societies, and all similar organizations, are doing great work for the future of our country. Birds should be saved because of utilitarian reasons; and, moreover, they should be saved because of reasons unconnected with any return in dollars and cents. A grove of giant redwoods or sequoias should be kept just as we keep a great and beautiful cathedral. The extermination of the passenger-pigeon meant that mankind was just so much poorer; exactly as in the case of the destruction of the cathedral at Rheims. And to lose the chance to see frigate-birds soaring in circles above the storm, or a file of pelicans winging their way homeward across the crimson afterglow of the sunset, or a myriad terns flashing in the bright light of midday as they hover in a shifting maze above the beach—why, the loss is like the loss of a gallery of the masterpieces of the artists of old time.

# 10

# Grand Canyon Speech, 1903

I HAVE COME HERE TO SEE THE GRAND CAÑON OF ARIZONA, because in that cañon Arizona has a natural wonder, which, so far as I know, is in its kind absolutely unparalleled throughout the rest of the world. I shall not attempt to describe it, because I cannot. I could not choose words that would convey or that could convey to any outsider what that cañon is. I want to ask you to do one thing in connection with it in your own interest, and in the interest of the country.

Keep this great wonder of nature as it now is.

I was delighted to learn of the wisdom of the Santa Fe Railroad in deciding not to build their hotel on the brink of the cañon. I hope you will not have a building of any kind, not a summer cottage, a hotel or anything else, to mar the wonderful grandeur, sublimity, the great loneliness and beauty of the cañon.

Leave it as it is. You cannot improve on it; not a bit. The ages have been at work on it, and man can only mar it. What you can do is to keep it for your children, your children's children and for all who come after you, as one of the great sights which every American, if he can travel at all, should see. Keep the Grand Cañon as it is.

PART 3

# NATURAL HISTORY

# 11

# Small Country Neighbors

SMALL MAMMALS, WITH THE EXCEPTION OF SQUIRRELS, ARE so much less conspicuous than birds, and indeed usually pass their lives in such seclusion, that the ordinary observer is hardly aware of the presence. At Sagamore Hill, for instance, except at haying time, I rarely see the swarming meadow-mice, the much less plentiful pine-mice, or the little mole-shrews, alive, unless they happen to drop into a pit or sunken area which has been dug at one point to let light through a window into the cellar. The much more graceful and attractive white-footed mice and jumping mice are almost as rarely seen, though if one does come across a jumping mouse it at once attracts attention by its extraordinary leaps. The jumping mouse hibernates, like the woodchuck and chipmunk. The other little animals just mentioned are abroad all winter, the meadow-mice under the snow, the white-footed mice, and often the shrews, above the snow. The telltale snow, showing all the tracks, betrays the hitherto unsuspected existence of many little creatures; and the commonest marks upon it are those of the rabbit and especially of the white-footed mouse. The shrew walks or trots and makes alternate footsteps in the snow. Whitefoot, on the contrary, always jumps, whether going slow or fast, and his hind feet leave their prints side by side, often with the mark where the tail has dragged. I think whitefoot is the most plentiful of all our furred wild creatures, taken as a whole. He climbs trees well. I have found his nest in an old vireo's nest; but more often under stumps or boards. The meadow-mice often live in the marshes and are entirely at home in the water.

The shrew-mouse which I most often find is a short-tailed, rather thick-set little creature, not wholly unlike his cousin the shrew-mole, and just as greedy and ferocious. When a boy I captured one of these mole-shrews and found to my astonishment that he was a blood-thirsty and formidable little beast of prey. He speedily killed and ate a partially grown white-footed mouse which I put in the same cage with him. (I think a full-grown mouse of this kind would be an overmatch for a shrew.) I then put a small snake in with him. The shrew was very active but seemed nearly blind, and as he ran to and fro he never seemed to be aware of the presence of anything living until he was close to it, when he would instantly spring on it like a tiger. On this occasion he attacked the little snake with great ferocity, and after an animated struggle in which the snake whipped and rolled all around the cage, throwing the shrew to and fro a dozen times, the latter killed and ate the snake in triumph. Larger snakes frequently eat shrews, by the way.

One of my boys—the special friend of Josiah the badger—once discovered a flying squirrel's nest, in connection with which a rather curious incident occurred. The little boy had climbed a tree which is hollow at the top; and in this hollow he discovered a flying-squirrel mother with six young ones. She seemed so tame and friendly that the little boy for a moment hardly realized that she was a wild thing, and called down that he had "found a guinea-pig up the tree." Finally the mother made up her mind to remove her family. She took each one in turn in her mouth and flew or sailed down from the top of the tree to the foot of another tree near by; ran up this, holding the little squirrel in her mouth; and again sailed down to the foot of another tree some distance off. Here she deposited her young one on the grass, and then, reversing the process, climbed and sailed back to the tree where the nest was; then she took out another young one and returned with it, in exactly the same fashion as with the first. She repeated this until all six of the young ones were laid on the bank, side by side in a row, all with their heads the same way. Finding that she was not molested she ultimately took all six of the little fellows back to her nest, where she reared her brood undisturbed.

Among the small mammals at Sagamore Hill the chipmunks are the most familiar and the most in evidence; for they readily become tame and confiding. For three or four years a chipmunk—I suppose the same chipmunk—has lived near the tennis-court; and it has developed the rather puzzling custom of sometimes scampering across the court while we are in the middle of a game. This has happened two or three times every year, and is rather difficult to explain, for the chipmunk could just as well go round the court, and there seems no possible reason why he should suddenly run out on it while the game is in full swing. If he is seen, every one stops to watch him, and then he may himself stop and sit up to look about; but we may not see him until just as he is finishing a frantic scurry across, in imminent danger of being stepped on.

Usually birds are very regular in their habits, so that not only the same species but the same individuals breed in the same places year after year. In spite of their wings they are almost as local as mammals, and the same pair will usually keep to the same immediate neighborhood, where they can always be looked for in their season. There are wooded or brush-grown swampy places not far from the White House where in the spring or summer I can count with certainty upon seeing wrens, chats, and the ground-loving Kentucky warbler; an attractive little bird, which, by the way, itself looks much like a miniature chat. There are other places, in the neighborhood of Rock Creek, where I can be almost certain of finding the blue-gray gnat-catcher, which ranks just next to the humming-bird itself in exquisite daintiness and delicacy. The few pairs of mocking-birds around Washington have just as sharply defined haunts.

Nevertheless, it is never possible to tell when one may run across a rare bird, and even birds that are not rare now and then show marked individual idiosyncrasy in turning up, or even breeding, in unexpected places. At Sagamore Hill, for instance, I never knew a purple finch to breed until the summer of 1906. Then two pairs nested with us, one right by the house and the other near the stable. My attention was drawn to them by the bold, cheerful singing of the males who were spurred to rivalry by one another's voices. In September of the same year, while sitting in a rocking-chair on the broad veranda and looking out over the Sound, I heard the unmistakable "ank-ank" of nuthatches from a young elm at one corner of the house. I strolled over, expecting to find the white-bellied nuthatch, which is rather common on Long Island. But instead there were a couple of red-bellied nuthatches, birds familiar to me in the northern woods, but which I had never before seen at Sagamore Hill. They were tame and fearless, running swiftly up and down the tree-trunk and around the limbs while I stood and looked at them not ten feet away. The two younger boys ran out to see them; and then we hunted up their picture in the Wilson. I find, by the way, that Audubon's and Wilson's are still the most satisfactory large ornithologies, at least for nature-lovers who are not specialists; but of course any attempt at serious study of our birds means recourse to the numerous and excellent books and pamphlets by recent observers.

In May, 1907, two pairs of robins built their substantial nests, and raised their broods, on the piazza at Sagamore Hill; one over the transom of the north hall door and one over the transom of the south hall door. Only one pair of purple finches returned to us this year; and for the first time in many years no Baltimore orioles built in the elm by the corner of the house; they began their nest, but for some reason left it unfinished. The red-winged blackbirds, however, were more plentiful than for years previously, and two pairs made their nests near the old barn, where the grass stood lush and tall; this was the first time they had ever built nearer than the wood-pile pond, and I believe it was owing to the season being so cold and wet. It was perhaps due to the same cause that so many black-throated green warblers spent June and July in the woods on our place; they must have been breeding, though I only noticed the males. Each kept to his own special tract of woodland, among the tops of the tall trees, seeming to prefer the locusts, and throughout June each sang all day long—a drawling, cadenced little warble of five or six notes, usually uttered at intervals of a few seconds; sometimes while the little bird was perched motionless, sometimes as it flitted and crawled actively among the branches. With the resident of one particular grove I became well acquainted, as I was chopping a path through the grove. Every day the little warbler was singing away in the grove when I reached it, one locust-tree being his favorite perch. He paid not the slightest attention to my chopping; whereas a pair of downy woodpeckers, and a pair of great crested flycatchers, both of which, evidently, were likewise nesting near by, were much put out by my presence. While listening to my little black-throated friend I would continually hear the songs of his cousins, the prairie-warbler, the redstart, the black-and-white creeper, and the Maryland yellowthroat, not to speak of other birds, towhees, oven-birds, thrashers, vireos, and the beautiful, golden-voiced wood-thrushes.

The black-throated green warbler has seemingly become a regular summer resident of Long Island, for after discovering them on my place I found that two or three bird-loving neighbors were already familiar with them, and I heard them on several different occasions as I rode through the country round about. I already knew as summer residents in my neighborhood the following representatives of the warbler family: The oven-bird, chat, black-and-white creeper, Maryland yellowthroat, summer yellowbird, prairie-warbler, pine-warbler, blue-winged warbler, golden-winged warbler (very rare), blue yellow-backed warbler, and redstart.

The black-throated green warbler as a breeder and summer resident is a newcomer who has extended his range southward. But this same summer I found one warbler, the presence of which, if more than accidental, means that a southern form is extending its range northward. This was the Dominican or yellow-throated warbler. Two of my bird-loving friends are Mrs. E. H. Swan, Jr., and Miss Alice Weeks. On July 4 Mrs. Swan told me that a new warbler, the yellow-throated, was living near their house, and that she and her husband had seen him on several occasions. I was rather skeptical and told her I thought that it must be a Maryland yellowthroat. Mrs. Swan meekly acquiesced in the theory that she might have been mistaken; but two or three days afterward she sent me word that she and Miss Weeks had seen the bird again, had examined it thoroughly through their glasses, and were sure it was a yellow-throated warbler. Accordingly, on the morning of the 8th, I walked down and met them both near Mrs. Swan's house, about a mile from Sagamore Hill. We did not have to wait long before we heard an unmistakably new warbler song—loud, ringing, sharply accented, just as the yellowthroat's song is described in Chapman's book. At first the little bird kept high in the tops of the pines, but after a while he came to the lower branches and we were able to see him distinctly. Only a glance was needed to show that my two friends were quite right in their identification, and that the bird was undoubtedly the Dominican or yellow-throated warbler. Its bill was as long as that of a black-and-white creeper, in sharp contrast to the bills of the other true wood-warblers, and the olive-gray back, yellow throat and breast, streaked sides, white belly, black cheek and forehead, and white line above eye and spot on the side of the neck, could all be plainly made out. The bird kept continually uttering its loud, sharply modulated and attractive warble. It never left the pines, and though continually on the move, it yet moved with a certain deliberation, like a pine-warbler and not with the fussy agility of most of its kinsfolk. Occasionally it would catch some insect on the wing, but most of the time kept hopping about among the pine-needles at the ends of the twig clusters, or moving along the larger branches, stopping from time to time to sing. Now and then it would sit still on one twig for several minutes, singing at short intervals and preening its feathers.

In one apple-tree we find a flicker's nest every year; the young make a queer, hissing, bubbling sound, a little like the boiling of a pot. This year one of the young ones fell out; I popped it back into the hole, whereupon its brothers and sisters "boiled" for several minutes, sounding like the caldron of a small and friendly witch. John Burroughs, and a Long Island neighbor, John Lewis Childs, came to see me one day, in June, 1907; and I was able to show them the various birds of most interest—the purple finch, the black-throated green warbler, the redwings in their unexpected nesting-place by the old barn, and the orchard-orioles and yellow-billed cuckoos in the garden.

At the White House we are apt to stroll around the grounds for a few minutes after breakfast; and during the migrations, especially in spring, I often take a pair of field-glasses so as to examine any bird as to the identity of which I am doubtful. From the end of April the warblers pass in troops—myrtle, magnolia, chestnut-sided, bay-breasted, blackburnian, black-throated blue, Canadian, and many others, with at the very end of the season the blackpolls; exquisite little birds, but not conspicuous as a rule, except perhaps the blackburnian, whose brilliant orange throat and breast flame when they catch the sunlight as he flits among the trees. The males in their dress of courtship are easily recognized by any one who has Chapman's book on the warblers. On May 4, 1906, I saw a Cape May warbler, the first I had ever seen. It was in a small pine. It was fearless, allowing a close approach, and as it was a male in high plumage, it was unmistakable.

In 1907, after a very hot week in early March, we had an exceedingly cold and late spring. The first bird I heard sing in the White House grounds was a white-throated sparrow on March 1, a song-sparrow speedily following. The whitethroats stayed with us until the middle of May, overlapping the arrival of the indigo-buntings; but during the last week in April and first week in May their singing was drowned by the music of the purple finches, which I never before saw in such numbers around the White House. When we sat by the south fountain, under an apple-tree then blossoming, sometimes three or four purple finches would be singing in the fragrant bloom overhead. In June a pair of wood-thrushes and a pair of black-and-white creepers made their homes in the White House grounds, in addition to our ordinary home-makers, the flickers, redheads, robins, catbirds, song-sparrows, chippies, summer yellowbirds, grackles, and I am sorry to say, crows. A handsome sapsucker spent a week with us. In this same year five night-herons spent January and February in a swampy tract by the Potomac, half a mile or so from the White House.

At Mount Vernon there are of course more birds than there are around the White House, for it is in the country. At present but one mocking-bird sings around the house itself, and in the gardens, and the woods of the immediate neighborhood. Phœbe-birds nest at the heads of the columns under the front portico; and a pair—or rather, doubtless, a succession of pairs—has nested in Washington's tomb itself, for the twenty years since I have known it. The cardinals, beautiful in plumage, and with clear ringing voices, are characteristic of the place. I am glad to say that the woods still hold many gray—not red—foxes, the descendants of those which Washington so perseveringly hunted.

At Oyster Bay on a desolate winter afternoon many years ago I shot an Ipswich sparrow on a strip of ice-rimmed beach, where the long coarse grass waved in front of a growth of blueberries, beach-plums, and stunted pines. I think it was the same winter that we were visited not only by flocks of crossbills, pine-linnets, redpolls, and pine-grosbeaks, but by a number of snowy owls, which flitted to and fro in ghostlike fashion across the wintry landscape and showed themselves far more diurnal in their habits than our native owls. One fall about the same time a pair of duck-hawks appeared off the bay. It was early, before many ducks had come, and they caused havoc among the night-herons, which were then very numerous in the marshes around Lloyd's Neck, there being a big heronry in the woods near by. Once I saw a duck-hawk come around the bend of the shore, and dart into a loose gang of young night-herons, still in the brown plumage, which had jumped from the marsh at my approach. The pirate struck down three herons in succession and sailed swiftly on without so much as looking back at his victims. The herons, which are usually rather dull birds, showed every sign of terror whenever the duck-hawk appeared in the distance; whereas, they paid no heed to the fish-hawks as they sailed overhead. The little fish-crows are not rare around Washington, though not so common as the ordinary crows; once I shot one at Oyster Bay. They are not so wary as their larger kinsfolk. The soaring turkey-buzzards, so beautiful on the wing and so loathsome near by, are seen everywhere around the Capital.

In Albemarle County, Virginia, we have a little place called Pine Knot, where we sometimes go, taking some or all of the children, for a three or four days' outing. It is a mile from the big stock farm "Plain Dealing," belonging to an old friend, Mr. Joseph Wilmer. The trees and flowers are like those of Washington, but their general close resemblance to those of Long Island is set off by certain exceptions. There are osage-orange hedges, and in spring many of the roads are bordered with bands of the brilliant yellow blossoms of the flowering broom, introduced by Jefferson. There are great willow-oaks here and there in the woods or pastures, and occasional groves of noble tulip-trees in the many stretches of forest; these trees growing to a much larger size than on Long Island. As at Washington, among the most plentiful flowers are the demure little Quaker ladies, which are not found at Sagamore Hill—where we also miss such northern forms as the wake-robin and the other trilliums, which used to be among the characteristic marks of springtime at Albany. At Pine

Knot the redbud, dogwood, and laurel are plentiful; though in the case of the last two no more so than at Sagamore Hill. The azalea—its Knickerbocker name in New York was pinkster—grows and flowers far more luxuriantly than on Long Island. The moccasin-flower and the china-blue Virginia cowslip with its pale-pink buds, the blood-red Indian pink, the painted columbine, and many, many other flowers somewhat less showy, carpet the woods. The birds are, of course, for the most part the same as on Long Island, but with some differences. These differences are, in part, due to the more southern locality; but in part I cannot explain them, for birds will often be absent from one place seemingly without any real reason. Thus around us in Albemarle County song-sparrows are certainly rare and I have not seen Savannah sparrows at all; but the other common sparrows, such as the chippy, field-sparrow, vesper-sparrow, and grasshopper-sparrow, abound; and in an open field, where bindweed, morning-glories, and evening primroses grew among the broom-sedge, I found some small grass-dwelling sparrows, which with the exercise of some little patience I was able to study at close quarters with the glasses; as I had no gun I could not be positive about their identification, though I was inclined to believe that they were Henslow's sparrows. Of birds of brilliant color there are six species—the cardinal, the summer redbird, and the scarlet tanager, in red; and the bluebird, indigo-bunting, and blue grosbeak, in blue. I saw but one pair of blue grosbeaks; but the little indigo-buntings abound, and bluebirds are exceedingly common, breeding in numbers. It has always been a puzzle to me why they do not breed around us at Sagamore Hill, where I only see them during the migrations. Neither the rosy summer redbirds nor the cardinals are quite as brilliant as the scarlet tanagers, which fairly burn like live flames; but the tanager is much less common than either of the others in Albemarle County, and it is much less common than it is at Sagamore Hill. Among the singers the wood-thrush is not common, but the meadow-lark abounds. The yellow-breasted chat is everywhere and in the spring its clucking, whistling, whooping, and calling seem never to stop for a minute. The white-eyed vireo is found in the same thick undergrowth as the chat, and among the smaller birds it is one of those most in evidence to the ear. In one or two place I came across parties of the long-tailed Bewick's wren, as familiar as the house-wren but with a very different song. There are gentle mourning-doves; and black-billed cuckoos seem more common than the yellowbills. The mocking-birds are, as always, most interesting. I was much amused to see one of them following two crows; when they lit in a ploughed field the mocking-bird paraded alongside of them six feet off, and then fluttered around to the attack. The crows, however, were evidently less bothered by it than they would have been by a king-bird. At Plain Dealing many birds nest within a stone's throw of the rambling, attractive house, with its numerous outbuildings, old garden, orchard, and venerable locusts and catalpas. Among them were Baltimore and orchard orioles, purple grackles, flickers, and red-headed woodpeckers, bluebirds, robins, king-birds, and indigo-buntings. One observation which I made was of real interest. On May 18, 1907, I saw a small party of a dozen or so passenger-pigeons, birds I had not seen for a quarter of a century and never expected to see again. I saw them two or three times flying hither and thither with great rapidity, and once they perched in a tall dead pine on the edge of an old field. They were unmistakable; yet the sight was so unexpected that I almost doubted my eyes, and I welcomed a bit of corroborative evidence coming from Dick, the colored foreman at Plain Dealing. Dick is a frequent companion of mine in rambles around the country, and he is an unusually close and accurate observer of birds, and of wild things generally. Dick had mentioned to me having seen some "wild carrier-pigeons," as he called them; and thinking over this remark of his after I had returned to Washington, I began to wonder whether he too might not have seen passenger-pigeons. Accordingly, I wrote to Mr. Wilmer, asking him to question Dick and find out what the "carrier-pigeons" looked like. His answering letter runs in part as follows:

> On May 12th last Dick saw a flock of about thirty wild pigeons, followed at a short distance by about half as many, flying in a circle very rapidly, between the Plain Dealing house and the woods, where they disappeared. They had pointed tails and resembled somewhat large doves—the breast and sides rather a brownish red. He had seen them before, but many years ago. I think it is unquestionably the passenger pigeon—*ectopistes*

> *migratoria*—described on page 25 of the fifth volume of Audubon. I remember the pigeon roosts as he describes them, on a smaller scale, but large flocks have not been seen in this part of Virginia for many years.

The house at Pine Knot consists of one long room, with a broad piazza, below, and three small bedrooms above. It is made of wood, with big outside chimneys at each end. Wood-rats and white-footed mice visit it; once a weasel came in after them; now a flying squirrel has made his home among the rafters. On one side the pines and on the other the oaks come up to the walls; in front the broom-sedge grows almost to the piazza, and above the line of its waving plumes we look across the beautiful rolling Virginia farm country to the foot-hills of the Blue Ridge. At night whippoorwills call incessantly around us. In the late spring or early summer we usually take breakfast and dinner on the veranda, listening to mocking-bird, cardinal, and Carolina wren, as well as to many more common singers. In the winter the little house can only be kept warm by roaring fires in the great open fireplaces, for there is no plaster on the walls, nothing but the bare wood. Then the table is set near the blazing logs at one end of the long room which makes up the lower part of the house, and at the other end the colored cook—Jim Crack by name—prepares the delicious Virginia dinner; while around him cluster the little darkies, who go on errands, bring in wood, or fetch water from the spring, to put in the bucket which stands below where the gourd hangs on the wall. Outside the wind moans or the still cold bites if the night is quiet; but inside there is warmth and light and cheer.

There are plenty of quail and rabbits in the fields and woods near by, so we live partly on what our guns bring in; and there are also wild turkeys. I spent the first three days of November, 1906, in a finally successful effort to kill a wild turkey. Each morning I left the house between three and five o'clock, under a cold, brilliant moon. The frost was heavy; and my horse shuffled over the frozen ruts as I rode after Dick. I was on the turkey grounds before the faintest streak of dawn had appeared in the east; and I worked as long as daylight lasted. It was interesting and attractive in spite of the cold. In the night we heard the quavering screech-owls;

and occasionally the hooting of one of their bigger brothers. At dawn we listened to the lusty hammering of the big log-cocks, or to the curious coughing or croaking sound of a hawk before it left its roost. Now and then loose flocks of small birds straggled by us as we sat in the blinds or rested to eat our lunch; chickadees, tufted tits, golden-crested kinglets, creepers, cardinals, various sparrows, and small woodpeckers. Once we saw a shrike pounce on a field-mouse by a haystack; once we came on a ruffed grouse sitting motionless in the road.

The last day I had with me Jim Bishop, a man who had hunted turkeys by profession, a hard-working farmer, whose ancestors have for generations been farmers and woodsmen; an excellent hunter, tireless, resourceful, with an eye that nothing escaped; just the kind of man one likes to regard as typical of what is best in American life. Until this day, and indeed until the very end of this day, chance did not favor us. We tried to get up to the turkeys on the roosts before daybreak; but they roosted in pines, and, night though it was, they were evidently on the lookout, for they always saw us long before we could make them out, and then we could hear them fly out of the tree-tops. Turkeys are quite as wary as deer, and we never got a sight of them while we were walking through the woods; but two or three times we flushed gangs, and my companion then at once built a little blind of pine boughs, in which we sat while he tried to call the scattered birds up to us by imitating, with marvelous fidelity, their yelping. Twice a turkey started toward us, but on each occasion the old hen began calling some distance off and all the scattered birds at once went toward her. At other times I would slip around to one side of a wood while my companion walked through it; but either there were no turkeys or they went out somewhere far away from us.

On the last day I was out thirteen hours. Finally, late in the afternoon, Jim Bishop marked a turkey into a point of pines which stretched from a line of wooded hills down into a narrow open valley on the other side of which again rose wooded hills. I ran down to the end of the point and stood behind a small oak, while Bishop and Dick walked down through the trees to drive the turkeys toward me. This time everything went well; the turkey came out of the cover not too far off and sprang into the air, heading across the valley and offering me a side shot at forty yards as he sailed by. It was just a distance for the close-shooting ten-bore duck gun I carried; and at the report down came the turkey in a heap, not so much as a leg or wing moving. It was an easy shot. But we had hunted hard for three days; and the turkey is the king of American game-birds; and besides I knew he would be very good eating indeed when we brought him home; so I was as pleased as possible when Dick lifted the fine young gobbler, his bronze plumage iridescent in the light of the westering sun.

Formerly we could ride across country in any direction around Washington; and almost as soon as we left the beautiful tree-shaded streets of the city we were in the real country. But as Washington grows, it naturally—and to me most regrettably—becomes less and less like its former, glorified-village, self; and wire fencing has destroyed our old cross-country rides. Fortunately there are now many delightful bridle trails in Rock Creek Park; and we have fixed up a number of good jumps at suitable places—a stone wall, a water jump, a bank with a ditch, two or three post-and-rails, about four feet high, and some stiff brush-hurdles, one of five feet seven inches. The last, which is the only formidable jump, was put up to please two sporting members of the administration, Bacon and Meyer. Both of them school their horses over it; and my two elder boys, and Fitzhugh Lee, my cavalry aide, also school my horses over it. On one of my horses, Roswell, I have gone over it myself; and as I weigh two hundred pounds without my saddle I think that the jump, with such a weight, in cold blood, should be credited to Roswell for righteousness. Roswell is a bay gelding; Audrey a black mare; they are Virginia horses. In the spring of 1907 I had photographs taken of them going over the various jumps. Roswell is a fine jumper, and usually goes at his jumps in a spirit of matter-of-fact enjoyment. But he now and then shows queer kinks in this temper. On one of these occasions he began by wishing to rush his jumps, and by trying to go over the wings instead of the jumps themselves. He fought hard for his head; and as it happened that the best picture we got of him in the air was at this particular time, it gives a wrong idea of his ordinary behavior, and also, I sincerely trust, a wrong idea of my hands. Generally he takes his jumps like a gentleman.

# 12

# The Wapiti or Round-Horned Elk

THE WAPITI IS THE LARGEST AND STATELIEST DEER IN THE world. A full-grown bull is a big as a steer. The antlers are the most magnificent trophies yielded by any game animal of America, save the giant Alaskan moose. When full grown they are normally of twelve tines; frequently the tines are more numerous, but the increase in their number has no necessary accompaniment in increase in the size of the antlers. The length, massiveness, roughness, spread, and symmetry of the antlers must all be taken into account in rating the value of a head. Antlers over fifty inches in length are large; if over sixty, they are gigantic. Good heads are getting steadily rarer under the persecution which has thinned out the herds.

Next to the bison the wapiti is of all the big-game animals of North America the one whose range has most decreased. Originally it was found from the Pacific coast east across the Alleghanies, through New York to the Adirondacks, through Pennsylvania into western New Jersey, and far down into the mid-country of Virginia and the Carolinas. It extended northward into Canada, from the Great Lakes to Vancouver; and southward into Mexico, along the Rockies. Its range thus corresponded roughly with that of the bison, except that it went farther west and not so far north. In the early colonial days so little heed was paid by writers to the teeming myriads of game that it is difficult to trace the wapiti's distribution in the Atlantic coast region. It was certainly killed out of the Adirondacks long before the period when the backwoodsmen were settling the valleys of the Alleghany Mountains; there they found the elk abundant, and the stately creatures roamed in great bands over Tennessee, Kentucky, Ohio, and Indiana when the first settlers made their way into what are now these States, at the outbreak of the Revolution. These first settlers were all hunters, and they followed the wapiti (or, as they always called it, the elk) with peculiar eagerness. In consequence its numbers were soon greatly thinned, and about the beginning of the present century it disappeared from that portion of its former range lying south of the Great Lakes and between the Alleghanies and the Mississippi. In the northern Alleghanies it held its own much longer, the last individual of which I have been able to get record having been killed in Pennsylvania in 1869. In the forests of northern Wisconsin, northern Michigan, and Minnesota wapiti existed still longer, and a very few individuals may still be found. A few are left in Manitoba. When Lewis and Clark and Pike became the pioneers among the explorers, army officers, hunters, and trappers who won for our people the great West, they found countless herds of wapiti throughout the high plains country from the Mississippi River to the Rocky Mountains. Throughout this region it was exterminated almost as rapidly as the bison, and by the early eighties there only remained a few scattered individuals, in bits of rough country such as the Black Hills, the sand-hills of Nebraska, and certain patches of Bad Lands along the Little Missouri. Doubtless stragglers exist even yet in one or two of these localities. But by the time the great buffalo-herds of the plains were completely exterminated, in 1883, the wapiti had likewise ceased to be plains animals; the peculiar Californian form had also been well-nigh exterminated.

The nature of its favorite haunts was the chief factor in causing it to suffer more than any other game in America, save the bison, from the persecution of hunters and settlers. The boundaries of its range have shrunk in far greater proportion than in the case of any of our other game animals, save only the great wild ox, with which it was once so commonly associated. The moose, a beast of the forest, and the caribou, which, save in the far North, is also a beast of the forest, have in most places greatly diminished in numbers, and have here and there been exterminated altogether from outlying portions of their range; but the wapiti, which, when free to choose, preferred to frequent the plains and open woods, has completely vanished from nine-tenths of the territory over which it roamed a century and a quarter ago. Although it was never found in any one place in such enormous numbers as the bison and the caribou, it nevertheless went in herds far larger than the herds of any other American game save the two mentioned, and was formerly very much more abundant within the area of its distribution than was the moose within the area of its distribution.

This splendid deer affords a good instance of the difficulty of deciding what name to use in treating of our American game. On the one hand, it is entirely undesirable to be pedantic; and on the other hand, it seems a pity, at a time when speech is written almost as much as spoken, to use terms which perpetually require explanation in order to avoid confusion. The wapiti is not properly an elk at all; the term wapiti is unexceptionable, and it is greatly to be desired that it should be generally adopted. But unfortunately it has not been generally adopted. From the time when our back-woodsmen first began to hunt the animal among the foot-hills of the Appalachian chains to the present day, it has been universally known as elk wherever it has been found. In ordinary speech it is never known as anything else, and only an occasional settler or hunter would understand what the word wapiti referred to. The book name is a great deal better than the common name; but, after all, it is only a book name. The case is almost exactly parallel to that of the buffalo, which was really a bison, but which lived as the buffalo, died as the buffalo, and left its name imprinted on our landscape as the buffalo. There is little use in trying to upset a name which is imprinted in our geography in hundreds of such titles as Elk Ridge, Elk Mountain, Elkhorn River. Yet in the books it is often necessary to call it the wapiti in order to distinguish it both from its differently named close kinsfolk of the Old World and from its more distant relatives with which it shares the name of elk.

Disregarding the Pacific coast form of Vancouver and the Olympian Mountains, the wapiti is now a beast of the Rocky Mountain region proper, especially in western Montana, Wyoming, and Colorado. Throughout these mountains its extermination, though less rapid than on the plains, has nevertheless gone on with melancholy steadiness. In the early nineties it was still as abundant as ever in large regions in western Wyoming and Montana and northwestern Colorado. In northwestern Colorado the herds are now represented by only a few hundred individuals. In western Montana they are scattered over a wider region and are protected by the denser timber, but are nowhere plentiful. They have nearly vanished from the Bighorn Mountains. They are still abundant in and around their great nursery and breeding-ground, the Yellowstone National Park. If this Park could be extended so as to take in part of the winter range to the south, it would help to preserve them, to the delight of all lovers of nature and to the great pecuniary benefit of the people of Wyoming and Montana. But at present the winter range south of the Park is filling up with settlers, and, unless the conditions change, those among the Yellowstone wapiti which would normally go south will more and more be compelled to winter among the mountains, which will mean such immense losses from starvation and deep snow that the southern herds will be wofully thinned.[1] Surely all men who care for nature, no less than all men who care for big-game hunting, should combine to try to see that not merely the States by the Federal authorities make every effort, and are given every power, to prevent the extermination of this stately and beautiful animal, the lordliest of the deer kind in the entire world.

The wapiti, like the bison, and even more than the whitetail deer, can thrive in widely varying surroundings. It is at home among the high mountains, in the deep forests, and on the treeless, level plains. It is rather omnivorous in its tastes, browsing and grazing on all kinds of trees, shrubs, and grasses. These traits and its hardihood make it comparatively easy to perpetuate in big parks and forest preserves in a semiwild condition; and it has thriven in such preserves and parks in many of the Eastern States. As it does not, by preference, dwell in such tangled forests as are the delight of the moose and the whitetail deer, it vanishes much quicker than either when settlers appear in the land. In the mountains and foot-hills its habitat is much the same as that of the mule-deer, the two animals being often found in the immediate neighborhood of each other. In such places the superior size and value of the wapiti put it at a disadvantage in the keen struggle for life, and when the rifle-bearing hunter appears upon the scene, it is killed out long before its smaller kinsman.

Moreover, the wapiti is undoubtedly subject to queer freaks of panic stupidity, or what seems like a mixture of tameness and of puzzled terror. At these times a herd will remain almost motionless, the individuals walking undecidedly to and fro, and neither flinching nor giving any other sign even when hit with a bullet. In the old days it was not uncommon for a professional hunter to destroy an entire herd of wapiti when one of these fits was on them. Even nowadays they sometimes behave this way. In 1897, Mr. Ansley Wilcox, of Buffalo, was hunting in the Teton basin. He came across a small herd of wapiti, the first he had ever seen, and opened fire when a hundred and fifty yards distant. They paid no heed to the shots, and after taking three or four at one bull, with seemingly no effect, he ran in closer and emptied his magazine at another, also seemingly without effect, before the herd slowly disappeared. After a few rods, both bulls fell; and on examination it was found that all nine bullets had hit them.

To my mind, the venison of the wapiti is, on the whole, better than that of any other wild game, though its fat, when cooled, at once hardens, like mutton-tallow.

In its life habits the wapiti differs somewhat from its smaller relatives. It is far more gregarious, and is highly polygamous. During the spring, while the bulls are growing their great antlers and while the cows have very young calves, both bulls and cows live alone, each individual for itself. At such time each seeks the most secluded situation, often going very high up on the mountains. Occasionally a couple of bulls lie together, moving around as little as possible. The cow at this time realizes that her calf's chance of life depends upon her absolute seclusion and avoids all observation.

As the horns begin to harden, the bulls thrash the velvet off against quaking asp or ash or even young spruce, splintering and battering the bushes and small trees. The cows and calves begin to assemble; the bulls seek them. But the bulls do not run the cows as among the smaller deer the bucks run the does. The time of the beginning of the rut varies in different places, but it usually takes place in September, about a month earlier than that of the deer in the same locality. The necks of the bulls swell and they challenge incessantly, for, unlike the smaller deer, they are very noisy. Their love and war calls, when heard at a little distance amid the mountains, have a most musical sound. Frontiersmen usually speak of their call as "whistling," which is not an appropriate term. The call may be given in a treble or in a bass, but usually consists of two or three bars, first rising and then falling, followed by a succession of grunts. The grunts can only be heard when close up. There can be no grander or more attractive chorus than the challenging of a number of wapiti bulls when two great herds happen to approach one another under the moonlight or in the early dawn. The pealing notes echo through the dark valleys as if from silver bugles, and the air is filled with the wild music. Where little molested the wapiti challenge all day long.

They can be easiest hunted during the rut, the hunter placing them and working up to them by the sound alone. The bulls are excessively truculent and pugnacious. Each big one gathers a herd of cows about him and drives all possible rivals away from his immediate neighborhood, although sometimes spike-bulls are allowed to remain with the herd. Where wapiti are very abundant, however, many of these herds may join together and become partially welded into a mass that may contain thousands of animals. In the old days such huge herds were far from uncommon, especially during the migrations; but nowadays there only remain one or two localities in which wapiti are sufficiently plentiful ever to come together in bands of any size. The bulls are incessantly challenging and fighting one another, and driving around the cows and calves. Each keeps the most jealous watch over his own harem, treating its members with great brutality, and is selfishly indifferent to their fate the instant he thinks his own life in jeopardy. During the rut the erotic manifestations of the bull are extraordinary.

One or two fawns are born about May. In the mountains the cow usually goes high up to bring forth her fawn. Personally I have only had a chance to observe the wapiti in spring in the neighborhood of my ranch in the Bad Lands of the Little Missouri. Here the cow invariably selected some wild, lonely bit of very broken country in which there were dense thickets and some water. There was one such patch some fifteen miles from my ranch, in which for many years wapiti regularly bred. The breeding cow lay by herself, although sometimes the young of the preceding year would lurk in the neighborhood. For the first few days the calf hardly left the bed, and would not move even when handled. Then it began to follow the mother. In this particular region the grass was coarse and rank, save for a few patches in the immediate neighborhood of the little alkali springs. Accordingly, it was not much visited by the cattle or by the cowboys. Doubtless in the happier days of the past, when man was merely an infrequent interloper, the wapiti cows had made their nurseries in pleasanter and more fruitful valleys. But in my time the hunted creatures had learned that their only chance was to escape observation. I have known not only cows with young calves, but cows when the calves were out of the spotted coat, and even yearlings, to try to escape by hiding—the great beasts lying like rabbits in some patch of thick brush while I rode close by. The best hunting-horse I ever had, old Manitou, in addition to his other useful qualities, would serve as a guard on such occasions. I would leave him on a little hillock to one side of such a patch of brush, and as he walked slowly about, grazing and rattling his bridle-chains, he would prevent the wapiti breaking cover on that side, and give me an additional chance of slipping around toward them—although if the animal was a cow, I never molested it unless in dire need of meat.

Most of my elk-hunting was done among the stupendous mountain masses of the Rockies, which I usually reached after a long journey, with wagon or pack-train, over the desolate plains. Ordinarily I planned to get to the hunting-ground by the end of August, so as to have ample time. By that date the calves were out of the spotted coat, the cows and the young of the preceding year had banded, and the big bulls had come down to join them from the remote recesses in which they had been lying, solitary or in couples, while their antlers were growing. Many bulls were found alone, or, if young, in small parties; but the normal arrangement was for each big bull to have his own harem, around the outskirts of which there were to be found lurking occasional spike-bulls or two-year-olds, who were always venturing too near and being chased off by the master-bull. Frequently several such herds joined together into a great band. Before the season was fairly on, when the bulls had not been worked into actual frenzy, there was not much fighting in these bands. Later they were the scenes of desperate combats. Each master bull strove to keep his harem under his own eyes, and was always threatening and fighting the other master-bulls, as well as those bulls whose prowess had proved insufficient hitherto to gain them a band, or who, after having gained one, had been so exhausted and weakened as to succumb to some new aspirant for the leadership. The bulls were calling and challenging all the time, and there was ceaseless turmoil, owing to their fights and their driving the cows around. The cows were more wary than the bulls, and there were so many keen noses and fairly good eyes that it was difficult to approach a herd; whereas the single bulls were so noisy, careless, and excited that it was comparatively easy to stalk them. A rutting wapiti bull is as wicked-looking a creature as can be imagined, swaggering among the cows and threatening young bulls, his jaws mouthing and working in a kind of ugly leer.

The bulls fight desperately with one another. The two combatants come together with a resounding clash of antlers, and then push and strain with their mouths open. The skin on their necks and shoulders is so thick and tough that the great prongs cannot get through or do more than inflict bruises. The only danger comes when the beaten party turns to flee. The victor pursues at full speed. Usually the beaten one gets off; but if by accident he is caught where he cannot escape, he is very apt to be gored in the flank and killed. Mr. Baillie-Grohman has given a very interesting description of one such fatal duel of which he was an eye-witness on a moonlight night in the mountains. I have never known of the bull trying to protect the cow from any enemy. He battles for her against rivals with intense ferocity; but his attitude toward her, once she is gained, is either that of brutality or of indifference. She will fight for her calf against any enemy which she thinks she has a chance of conquering, although of course not against man. But the bull leaves his family to their fate the minute he thinks there is any real danger. During the rut he is greatly excited, and does not fear a dog or a single wolf, and may join with the rest of the herd of both sexes in trying to chase off one or the other, should he become aware of its approach. But if there is serious danger, his only thought is for himself, and he has no compunctions about sacrificing any of his family. When on the move a cow almost always goes first, while the bull brings up the rear.

In domestication the bulls are very dangerous to human beings, and will kill a man at once if they can get him at a disadvantage; but in a state of nature they rarely indeed overcome their abject terror of humanity, even when wounded and cornered. Of course, if the man comes straight up to him where he cannot get away, a wapiti will fight as, under like circumstances, a blacktail or whitetail will fight, and equally, of course, he is then far more dangerous than his smaller kinsfolk; but he is not nearly so apt to charge as a bull moose. I have never known but two authentic instances of their thus charging. One happened to a hunter named Bennett, on the Little Missouri; the other to a gentleman I met, a doctor, in Meeker, Col. The doctor had wounded his wapiti, and as it was in the late fall, followed him easily in the snow. Finally he came upon the wapiti standing where the snow was very deep at the bottom of a small valley, and on his approach the wapiti deliberately started to break his way through the snow toward him, and had almost reached him when he was killed. But for every one such instance of a wapiti's charging there are a hundred in which a bull moose has charged. Senator Redfield Proctor was charged most resolutely by a mortally hurt bull moose which fell in death-throes just before reaching him; and I could cite case after case of the kind.

The wapiti's natural gaits are a walk and a trot. It walks very fast indeed, especially if travelling to reach some given point. More than once I have sought to overtake a travelling bull, and have found myself absolutely unable to do so, although it never broke its walk. Of course, if I had not been obliged to pay any heed to cover or wind, I could have run up on it; but the necessity for paying heed to both handicapped me so that I was actually unable to come up to the quarry as it swung steadily on through woodland and open, over rough ground and smooth. Wapiti have a slashing trot, which they can keep up for an indefinite time and over any kind of country. Only a good pony can overtake them when they have had any start and have settled into this trot. If much startled they break into a gallop—the young being always much more willing to gallop than the old. Their gallop is very fast, especially downhill.

But they speedily tire under it, A yearling or a two-year-old can keep it up for a couple of miles. A heavy old bull will be done out after a few hundred yards. I once saw a band of wapiti frightened into a gallop down a steep incline where there were also a couple of mule-deer. I had not supposed that wapiti ran as fast as mule-deer, but this particular band actually passed the deer, though the latter were evidently doing their best; the wapiti were well ahead, when, after thundering down the steep, broken incline, they all disappeared into a belt of woodland. In spite of their size, wapiti climb well and go surefootedly over difficult and dangerous ground. They have a bad habit of coming out to the edges of cliffs, or on mountain spurs, and looking over the landscape beneath, almost as though they enjoyed the scenery. What their real object is on such occasions I do not know.

The nose of the wapiti is very keen. Its sight is much inferior to that of the antelope, but about as good as a deer's. Its hearing is also much like that of a deer. When in country where it is little molested, it feeds and moves about freely by day, lying down to rest at intervals, like cattle. Wapiti offer especial attractions to the hunter, and next to the bison are more quickly exterminated than any other kind of game. Only the fact that they possess a far wider range of habitat than either the mule-deer, the prongbuck, or the moose has enabled them still to exist. Their gregariousness is also against them. Even after the rut the herds continue together until in midspring the bulls shed their antlers—for they keep their antlers at least two months longer than deer. During the fall, winter, and early spring wapiti are roving, restless creatures. Their habit of migration varies with locality, as among mule-deer. Along the Little Missouri, as in the plains country generally, there was no well-defined migration. Up to the early eighties, when wapiti were still plentiful, the bands wandered far and wide, but fitfully and irregularly, wholly without regard to the season, save that they were stationary from May to August. After 1883 there were but a few individuals left, although as late as 1886 I once came across a herd of nine. These surviving individuals had learned caution. The bulls only called by night, and not very frequently then, and they spent the entire year in the roughest and most out-of-the-way places, having the same range both winter and summer. They selected tracts where the ground was very broken and there were much shrubbery and patches of small trees. This tree and bush growth gave them both shelter and food; for they are particularly fond of browsing on the leaves and tender twig ends, though they also eat weeds and grass.

Wherever wapiti dwell among the mountains they make regular seasonal migrations. In northwestern Wyoming they spend the summer in the Yellowstone National Park, but in winter some go south to Jackson's Hole, while others winter in the park to the northeast. In northwestern Colorado their migrations followed much the same line as those of the mule-deer. In different localities the length of the migration, and even the time, differed. There were some places where the shift was simply from the high mountains down to their foot-hills. In other places great herds travelled a couple of hundred miles, so that localities absolutely barren one month would be swarming with wapiti the next. In some places the shift took place as early as the month of August; in others not until after the rut, in October or even November; and in some places the rut took place during the migration.

No chase is more fascinating than that of the wapiti. In the old days, when the mighty-antlered beasts were found upon the open plains, they could be followed upon horseback, with or without hounds. Nowadays, when they dwell in the mountains, they are to be killed only by the rifle-bearing still-hunter. Needless butchery of any kind of animal is repulsive, but in the case of the wapiti it is little short of criminal. He is the grandest of the deer kind throughout the world, and he has already vanished from most of the places where he once dwelt in his pride. Every true sportsman should feel it incumbent upon him to do all in his power to preserve so noble a beast of the chase from extinction. No harm whatever comes to the species from killing a certain number of bulls; but an excessive number should never be killed, and no cow or calf should under any circumstances be touched. Formerly, when wapiti were plentiful, it would have been folly for hunters and settlers in the unexplored wilderness not to kill wild game for their meat, and occasionally a cow or a calf had to be thus slain; but there is no excuse nowadays for a hunting-party killing anything but a full-grown bull.

In a civilized and cultivated country wild animals only continue to exist at all when preserved by sportsmen. The excellent people who protest against all hunting, and consider sportsmen as enemies of wild life, are ignorant of the fact that in reality the genuine sportsman is by all odds the most important factor in keeping the larger and more valuable wild creatures from total extermination. Of course, if wild animals were allowed to breed unchecked, they would, in an incredibly short space of time, render any country uninhabitable by man—a fact which ought to be a matter of elementary knowledge in any community where the average intelligence is above that of certain portions of Hindoostan. Equally, of course, in a purely utilitarian community all wild animals are exterminated out of hand. In order to preserve the wild life of the wilderness at all, some middle ground must be found between brutal and senseless slaughter and the unhealthy sentimentalism which would just as surely defeat its own end by bringing about the eventual total extinction of the game. It is impossible to preserve the larger wild animals in regions thoroughly fit for agriculture; and it is perhaps too much to hope that the larger carnivora can be preserved for merely æsthetic reasons. But throughout our country there are large regions entirely unsuited for agriculture, where, if the people only have foresight, they can, through the power of the State, keep the game in perpetuity. There is no hope of preserving the bison permanently, save in large private parks; but all other game, including not merely deer, but the pronghorn, the splendid bighorn, and the stately and beautiful wapiti, can be kept on the public lands, if only the proper laws are passed and if only these laws are properly enforced.

Most of us, as we grow older, grow to care relatively less for sport than for the splendid freedom and abounding health of outdoor life in the woods, on the plains, and among the great mountains; and to the true nature-lover it is melancholy to see the wilderness stripped of the wild creatures which gave it no small part of its peculiar charm. It is inevitable and probably necessary that the wolf and the cougar should go; but the bighorn and white goat among the rocks, the blacktail and wapiti grouped on the mountainside, the whitetail and moose feeding in the sedgy ponds—these add beyond measure to the wilderness landscape, and if they are taken away they leave a lack which nothing else can quite make good. So it is of those true birds of the wilderness, the eagle and the raven, and, indeed, of all the wild things, furred, feathered, and finned.

A peculiar charm in the chase of the wapiti comes from the wild beauty of the country in which it dwells. The moose lives in marshy forests; if one would seek the white goat or caribou of the northern Rockies, he must travel on foot, pack on back; while the successful chase of the bighorn, perhaps on the whole the manliest of all our sports, means heart-breaking fatigue for any but the strongest and hardiest. The prongbuck, again, must be followed on the desolate, sun-scorched plains. But the wapiti now dwells amid lofty, pine-clad mountains, in a region of lakes and streams. A man can travel in comfort while hunting it, because he can almost always take a pack-train with him, and the country is usually sufficiently open to enable the hunter to enjoy all the charm of the distant landscapes. Where the wapiti lives the spotted trout swarm in the brooks, and the wood-grouse fly upward to perch among the tree-tops as the hunter passes them. When hunting him there is always sweet cold water to be drunk at night, and beds of aromatic fir boughs on which to sleep with the blankets drawn over one to keep out the touch of the frost. He must be followed on foot, and the man who follows him must be sound in limb and wind. But his pursuit does not normally mean such wearing exhaustion as is entailed by climbing cliffs all day long after the white goat. Whoever has hunted the wapiti, as he looks at his trophies will always think of the great mountains with the snow lying in the rifts in their sides; of the splashing murmur of rock-choked torrents; of the odorous breath of the pine branches; of tents pitched in open glades; of long walks through cool, open forests; and of the great campfires, where the pitchy stumps flame like giant torches in the darkness.

In the old days, of course, much of the hunting was done on the open plains or among low, rugged hills. The wapiti that I shot when living at my Little Missouri ranch were killed under exactly the same conditions as mule-deer. When I built my ranch-house, wapiti were still not uncommon, and their shed antlers were very numerous both on the bottoms and in places among the hills. There was one such place a couple of miles from my ranch in a stretch of comparatively barren but very broken hill-country in which there were many score of these shed antlers. Evidently a few years before this had been a great gathering-place for wapiti toward the end of winter. My ranch itself derived its name, "The Elkhorn," from the fact that on the ground where we built it were found the skulls and interlocked antlers of two wapiti bulls who had perished from getting their antlers fastened in a battle. I never, however, killed a wapiti while on a day's hunt from the ranch itself. Those that I killed were obtained on regular expeditions, when I took the wagon and drove off to spend a night or two on ground too far for me to hunt it through in a single day from the ranch. Moreover, the wapiti on the Little Missouri had been so hunted that they had entirely abandoned the diurnal habits of their kind, and it was a great advantage to get on the ground early. This hunting was not carried on amid the glorious mountain scenery which marks the home of the wapiti in the Rockies; but the surroundings had a charm of their own. All really wild scenery is attractive. The true hunter, the true lover of the wilderness, loves all parts of the wilderness, just as the true lover of nature loves all seasons. There is no season of the year when the country is not more attractive than the city; and there is no portion of the wilderness where game is found in which it is not a keen pleasure to hunt. Perhaps no other kind of country quite equals that where snow lies on the lofty mountain peaks, where there are many open glades in the pine forests, and clear mountain lakes and rushing trout-filled torrents. But the fantastic desolation of the Bad Lands, and the endless sweep of the brown prairies, alike have their fascination for the true lover of nature and lover of the wilderness who goes through them on foot or on horseback. As for the broken hill-country in which I followed the wapiti and the mule-deer along the Little Missouri, it would be strange indeed if any one found it otherwise than attractive in the bright, sharp fall weather. Long, grassy valleys wound among the boldly shaped hills. The basins were filled with wind-beaten trees and brush, which generally also ran alongside of the dry watercourses down the middle of each valley. Cedars clustered in the sheer ravines, and here and there groups of elm and ash grew to a considerable height in the more sheltered places. At the first touch of the frost foliage turned russet or yellow—the Virginia creepers crimson. Under the cloudless blue sky the air was fresh and cool, and as we lay by the camp-fire at night the stars shone with extraordinary brilliancy. Under such conditions the actual chase of the wapiti was much like that of the mule-deer. They had been so hunted that they showed none of the foolish traits which they are prone to exhibit when bands are found in regions where they have been little persecuted; and they were easier to kill than mule-deer simply because they were more readily tracked and more readily seen, and offered a larger and on the whole a steadier, mark at which to shoot. When a small band had visited a pool their tracks could be identified at once, because in the soft ground the flexible feet spread and yielded so as to leave the marks of the false hoofs. On ordinary ground it was difficult to tell their footprints from those of the yearling and two-year-old ranch cattle.

But the mountains are the true ground for the wapiti. Here he must be hunted on foot, and nowadays, since he has grown wiser, skill and patience and the capacity to endure fatigue and exposure, must be shown by the successful hunter. My own wapiti-hunting has been done in September and early October, during the height of the rut, and therefore at a time when the conditions were most favorable for the hunter. I have hunted them in many places throughout the Rockies, from the Bighorn, in western Wyoming, to the Big Hole Basin, in western Montana, close to the Idaho line. Where I hunted, the wapiti were always very noisy both by day and by night, and at least half of the bulls that I killed attracted my attention by their calling before I saw either them or their tracks. At night they frequently passed close to camp, or came nearly up to the picketed horses, challenging all the time. More than once I slipped out, hoping to kill one by moonlight, but I never succeeded. Occasionally, when they were plentiful, and were restless and always roving about, I simply sat still on a log until one gave me a chance. Sometimes I came across them while hunting through likely localities, going up or across wind, keeping the sharpest lookout and moving with great care and caution, until I happened to strike the animals I was after. More than once I took the trail of a band, when out with some first-class woodsman, and after much running, dodging, and slipping through the timber, overtook the animals—though usually when thus merely following the trail

I failed to come up with them. On two different occasions I followed and came up to bands, attracted by their scent. Wapiti have a strong, and, on the whole, pleasing scent, like that of Alderney cattle, although in old bulls it becomes offensively strong. This scent is very penetrating. I once smelt a herd which was lying quite still taking its noonday siesta, certainly half a mile to the windward of me; and creeping up I shot a good bull as he lay. On another occasion, while working through the tangled trees and underbrush at the bottom of a little winding valley, I suddenly smelt wapiti ahead, and without paying any further attention to the search for tracks, I hunted cautiously up the valley, and when it forked was able to decide by the smell alone which way the wapiti had gone. He was going up-wind ahead of me, and his ground-covering walk kept me at a trot in order to overtake him. Finally I saw him before he saw me, and then, by making a run to one side, got a shot at him when he broke cover, and dropped him.

It is exciting to creep up to a calling wapiti. If it is a solitary bull, he is apt to be travelling, seeking the cows or on the lookout for some rival of weaker thews. Under such circumstances only hard running will enable the hunter to overtake him, unless there is a chance to cut him off. If, however, he hears another bull or has a herd under him, the chances are that he is nearly stationary or at least is moving slowly, and the hunter has every opportunity to approach. Of course, if there has been much hunting, even such a bull is wary and is on the lookout for harm. But in remote localities he becomes so absorbed in finding out the whereabouts of his rival, and is so busy answering the latter's challenges and going through motions of defiance, that with proper care it is comparatively easy to approach him. Once, when within seventy yards of such a bull, he partly made me out and started toward me. Evidently he could not tell exactly what I was—my buckskin shirt probably helping to puzzle him—and in his anger and eagerness he did not think of danger until it was too late. On another occasion I got up to two bulls that were fighting, and killed both. In the fights, weight of body seems to count for more than size of antlers.

Once I spent the better part of a day in following a wapiti bull before I finally got him. Generally when hunting wapiti I have been with either one of my men from the ranch or a hunter like Tazewell Woody, or John Willis. On this particular occasion however, I happened to be alone; and though I have rarely been as successful alone as when in the company of some thoroughly trained and experienced plainsman or mountain-man, yet when success does come under such circumstances it is always a matter of peculiar pride.

At the time, I was camped in a beautiful valley among the mountains which divide southwestern Montana from Idaho. The weather was cold and there were a couple of inches of snow on the ground, so that the conditions were favorable for tracking and stalking. The country was well wooded, but the forest was not dense, and there were many open glades. Early one morning, just about dawn, the cook, who had been up for a few minutes, waked me, to say that a bull wapiti was calling not far off. I rolled out of my bed and was dressed in short order. The bull had by this time passed the camp, and was travelling toward a range of mountains on the other side of the stream which ran down the valley bottom. He was evidently not alarmed, for he was still challenging. I gulped down a cup of hot coffee, munched a piece of hardtack, and thrust four or five other pieces and a cold elk tongue into my hunting-shirt, and then, as it had grown light enough to travel, started after the wapiti. I supposed that in a few minutes I should either have overtaken him or abandoned the pursuit, and I took the food with me simply because in the wilderness it never pays to be unprepared for emergencies. The wisdom of such a course was shown in this instance by the fact that I did not see camp again until long after dark.

I at first tried to cut off the wapiti by trotting through the woods toward the pass for which I supposed he was headed. The morning was cold, and, as always happens at the outset when one starts to take violent exercise under such circumstances, the running caused me to break into a perspiration; so that the first time I stopped to listen for the wapiti a regular fog rose over my glasses and then froze on them. I could not see a thing, and after wiping them found I had to keep gently moving in order to prevent them from clouding over again. It is on such cold mornings, or else in very rainy weather, that the man who has not been gifted with good eyes is most sensible of his limitations. I once lost a caribou which I had been following at speed over the snow because when I came into sight and halted, the moisture instantly formed and froze on my glasses so that I could not see anything, and before I got them clear the game had vanished. Whatever happened, I was bound that I should not lose this wapiti from a similar accident.

However, when I next heard him he had evidently changed his course and was going straight away from me. The sun had now risen, and following after him I soon found his tracks. He was walking forward with the regular wapiti stride, and I made up my mind I had a long chase ahead of me. We were going uphill, and though I walked hard, I did not trot until we topped the crest. Then I jogged along at a good gait, and as I had on moccasins and the woods were open, I did not have to exercise much caution. Accordingly I gained, and felt I was about to come up with him, when the wind brought down from very far off another challenge. My bull heard it before I did, and instantly started toward the spot at a trot. There was not the slightest use of my attempting to keep up with this, and I settled down into a walk. Half an hour afterward I came over a slight crest, and immediately saw a herd of wapiti ahead of me, across the valley on an open hillside. The herd was in commotion, the master-bull whistling vigorously and rounding up his cows, evidently much excited at the new bull's approach. There were two or three yearlings and two-year-old bulls on the outskirts of the herd, and the master-bull, whose temper had evidently not been improved by the coming of the stranger, occasionally charged these and sent them rattling off through the bushes. The ground was so open between me and them that I dared not venture across it, and I was forced to lie still and await developments. The bull I had been following and the herd bull kept challenging vigorously, but the former probably recognized in the latter a heavier animal, and could not rouse his courage to the point of actually approaching and doing battle. It by no means follows that the animal with the heaviest body has the best antlers, but the hesitation thus shown by the bull I was following made me feel that the other would probably yield the more valuable trophies, and after a couple of hours I made up my mind to try and get near the herd, abandoning the animal I had been after.

The herd showed but little symptoms of moving, the cows when let alone scattering out to graze, and some of them even lying down. Accordingly I did not hurry myself, and spent considerably over an hour in slipping off to the right and approaching through a belt of small firs. Unfortunately, however, the wind had slightly shifted, and while I was out of sight of the herd they had also come down toward the spot from whence I had been watching them. Accordingly, just as I was beginning to creep forward with the utmost caution, expecting to see them at any moment, I heard a thumping and cracking of branches that showed they were on the run. With wapiti there is always a chance of overtaking them after they have first started, because they tack and veer and halt to look around. Therefore I ran forward as fast as I could through the woods; but when I came to the edge of the fir belt I saw that the herd were several hundred yards off. They were clustered together and looking back, and saw me at once.

Off they started again. The old bull, however, had neither seen me nor smelt me, and when I heard his whistle of rage I knew he had misinterpreted the reason for the departure of his cows, and in another moment he came in sight, evidently bent on rounding them up. On his way he attacked and drove off one of the yearlings, and then took after the cows, while the yearling ran toward the outlying bull. The latter evidently failed to understand what had happened; at least he showed no signs of alarm. Neither, however, did he attempt to follow the fleeing herd, but started off again on his own line.

I was sure the herd would not stop for some miles, and accordingly I resumed my chase of the single bull. He walked for certainly three miles before he again halted, and I was then half a mile behind him. On this occasion he struck a small belt of woodland and began to travel to and fro through it, probably with an idea of lying down. I was able to get up fairly close by crawling on all fours through the snow for part of the distance; but just as I was about to fire he moved slightly, and though my shot hit him, it went a little too far back. He plunged over the hill crest and was off at a gallop, and, after running forward and failing to overtake him in the first rush, I sat down to consider matters. The snow had begun to melt under the sun, and my knees and the lower parts of my sleeves were wet from my crawl, and I was tired and hungry and very angry at having failed to kill the wapiti. It was, however, early in the afternoon, and I thought that if I let the wapiti alone for an hour, he would lie down, and then grow stiff and reluctant to get up; while in the snow I was sure I could easily follow his tracks. Therefore I ate my lunch, and then swallowed some mouthfuls of snow in lieu of drinking.

An hour afterward I took the trail. It was evident the bull was hard hit, but even after he had changed his plunging gallop for a trot he showed no signs of stopping; fortunately his trail did not cross any other. The blood signs grew infrequent, and two or three times he went up places which made it difficult for me to believe he was much hurt. At last, however, I came to where he had lain down; but he had risen again and gone forward. For a moment I feared that my approach had alarmed him, but this was evidently not the case, for he was now walking. I left the trail, and turning to one side below the wind I took a long circle and again struck back to the bottom of the valley down which the wapiti had been travelling. The timber here was quite thick, and I moved very cautiously, continually halting and listening for five or ten minutes. Not a sound did I hear, and I crossed the valley bottom and began to ascend the other side without finding the trail. Unless he had turned off up the mountains I knew that this meant he must have lain down; so I retraced my steps and with extreme caution began to make my way up the valley. Finally I came to a little opening, and after peering about for five minutes I stepped forward, and instantly heard a struggling and crashing in a clump of young spruce on the other side. It was the wapiti trying to get on his feet. I ran forward at my best pace, and as he was stiff and slow in his movements I was within seventy yards before he got fairly under way. Dropping on one knee, I fired and hit him in the flank. At the moment I could not tell whether or not I had missed him, for he gave no sign; but, running forward very fast, I speedily saw him standing with his head down. He heard me and again started, but at the third bullet down he went in his tracks, the antlers clattering loudly on the branches of a dead tree.

The snow was melting fast, and for fear it might go off entirely, so that I could not follow my back track, I went up the hillside upon which the wapiti lay, and taking a dead tree dragged it down to the bottom, leaving a long furrow. I then repeated the operation on the opposite hillside, thus making a trace which it was impossible for any one coming up or down the valley to overlook; and having conned certain landmarks by which the valley itself could be identified, I struck toward camp at a round trot; for I knew that if I did not get into the valley where the tent lay before dark, I should have to pass the night out. However, the last uncertain light of dusk just enabled me to get over a spur from which I could catch a glimpse of the camp-fire, and as I stumbled toward it through the forest I heard a couple shots, which showed that the cook and packer were getting anxious as to my whereabouts.

# 13

# On the Cattle Ranges

EARLY ONE JUNE JUST AFTER THE CLOSE OF THE REGULAR spring round-up, a couple of wagons with a score of riders between them were sent to work some hitherto untouched country between the Little Missouri and the Yellowstone. I was to go as the representative of our own and of one or two neighboring brands; but as the round-up had halted near my ranch I determined to spend a day there and then to join the wagons—the appointed meeting-place being a cluster of red scoria buttes, some forty miles distant, where there was a spring of good water.

Most of my day at the ranch was spent in slumber; for I had been several weeks on the round-up, where nobody ever gets quite enough sleep. This is the only drawback to the work; otherwise it is pleasant and exciting, with just that slight touch of danger necessary to give it zest, and without the wearing fatigue of such labor as lumbering or mining. But there is never enough sleep, at least on the spring and midsummer round-ups. The men are in the saddle from dawn until dusk, at the time when the days are longest on these great northern plains; and in addition there is the regular night guarding and now and then a furious storm or a stampede, when for twenty hours at a stretch the riders only dismount to change horses or snatch a mouthful of food.

I started in the bright sunrise, riding one horse and driving loose before me eight others, one carrying my bedding. They travelled strung out in a single file. I kept them trotting and loping, for loose horses are easiest to handle when driven at some speed, and moreover the way was long. My rifle was slung under my thigh; the lariat was looped on the saddle-horn.

At first our trail led through winding coulées and sharp grassy defiles; the air was wonderfully clear, the flowers were in bloom, the breath of the wind in my face was odorous and sweet. The patter and beat of the unshod hoofs rising in half-rhythmic measure frightened the scudding deer; but the yellow-breasted meadow-larks, perched on the budding tops of the bushes, sang their rich, full songs without heeding us as we went by.

When the sun was well on high and the heat of the day had begun we came to a dreary and barren plain, broken by rows of low clay buttes. The ground in places was whitened by alkali; elsewhere it was dull gray. Here there grew nothing save sparse tufts of coarse grass, and cactus, and sprawling sage-brush. In the hot air all things seen afar danced and wavered. As I rode and gazed at the shimmering haze the vast desolation of the landscape bore on me; it seemed as if the unseen and unknown powers of the wastes were moving by and marshaling their silent forces. No man save the wilderness-dweller knows the strong melancholy fascination of these long rides through lonely lands.

At noon, that the horses might graze and drink, I halted where some box-alders grew by a pool in the bed of a half-dry creek; and shifted my saddle to a fresh beast. When we started again we came out on the rolling prairie, where the green sea of wind-rippled grass stretched limitless as far as the eye could reach. Little striped gophers scuttled away, or stood perfectly straight at the mouths of their burrows, looking like picket-pins. Curlews clamored mournfully as they circled overhead. Prairie-fowl swept off clucking and calling, or strutted about with their sharp tails erect. Antelope were very plentiful, running like race-horses across the level, or uttering their queer, barking grunt as they stood at gaze, the white hairs on their rumps all on end, their neck-bands of broken brown and white vivid in the sunlight. They were found singly or in small straggling parties; the master-bucks had not yet begun to drive out the younger and weaker ones, as later in the season, when each would gather into a herd as many does as his jealous strength could guard from rivals. The nursing does whose kids had come early were often found with the bands; the others kept apart. The kids were very conspicuous figures on the prairies, across which they scudded like jack-rabbits, showing nearly as much speed and alertness as their parents; only the very young sought safety by lying flat to escape notice.

The horses cantered and trotted steadily over the mat of buffalo-grass, steering for the group of low scoria mounds which was my goal. In mid-afternoon I reached it. The two wagons were drawn up near the spring; under them lay the night wranglers, asleep; near by the teamster cooks were busy about the evening meal. A little way off the two day wranglers were watching the horse-herd; into which I speedily turned my own animals. The riders had already driven in the bunches of cattle, and were engaged in branding the calves and turning loose the animals that were not needed, while the remainder were kept, forming the nucleus of the herd which was to accompany the wagon.

As soon as the work was over the men rode to the wagons; sinewy fellows with tattered broad-brimmed hats and clanking spurs, some wearing leather chaps or leggings, others having their trousers tucked into their high-heeled top-boots, all with their flannel shirts and loose neckerchiefs dusty and sweaty. A few were indulging in rough, good-natured horse-play, to an accompaniment of yelling mirth; most were grave and taciturn, greeting me with a silent nod or a "How! friend." A very talkative man, unless the acknowledged wit of the party, according to the somewhat florid frontier notion of wit, is always looked on with disfavor in a cow camp. After supper, eaten in silent haste, we gathered round the embers of the small fires, and the conversation glanced fitfully over the threadbare subjects common to all such camps; the antics of some particularly vicious bucking bronco, how the different brands of cattle were showing up, the smallness of the calf drop, the respective merits of rawhide lariats and grass ropes, and bits of rather startling and violent news concerning the fates of certain neighbors. Then one by one we began to turn in under our blankets.

Our wagon was to furnish the night guards for the cattle; and each of us had his gentlest horse tied ready to hand. The night guards went on duty two at a time for two-hour watches. By good luck my watch came last. My comrade was a happy-go-lucky young Texan who for some inscrutable reason was known as "Latigo Strap"; he had just come from the South with a big drove of trail cattle.

A few minutes before two, one of the guards who had gone on duty at midnight rode into camp and wakened us up by shaking our shoulders. Fumbling in the dark, I speedily saddled my horse; Latigo had left his saddled and he started ahead of me. One of the annoyances of night guarding, at least in thick weather, is the occasional difficulty of finding the herd after leaving camp, or in returning to camp after the watch is over; there are few things more exasperating than to be helplessly wandering about in the dark under such circumstances. However, on this occasion there was no such trouble; for it was a brilliant starlight night and the herd had been bedded down by a sugar-loaf butte which made a good landmark. As we reached the spot we could make out the loom of the cattle lying close together on the level plain; and then the dim figure of a horseman rose vaguely from the darkness and moved by in silence; it was the other of the two midnight guards, on his way back to his broken slumber.

At once we began to ride slowly round the cattle in opposite directions. We were silent, for the night was clear and the herd was quiet; in wild weather, when the cattle are restless, the cowboys never cease calling and singing as they circle them, for the sounds seem to quiet the beasts.

For over an hour we steadily paced the endless round, saying nothing, with our greatcoats buttoned, for the air was chill toward morning on the northern plains, even in summer. Then faint streaks of gray appeared in the east. Latigo Strap began to call merrily to the cattle. A coyote came sneaking over the butte near by and halted to yell and wail; afterward he crossed the coulée and from the hillside opposite again shrieked in dismal crescendo. The dawn brightened rapidly; the little skylarks of the plains began to sing, soaring far overhead, while it was still much too dark to see them. Their song is not powerful, but it is so clear and fresh and long-continued that it always appeals to one very strongly; especially because it is most often heard in the rose-tinted air of the glorious mornings, while the listener sits in the saddle, looking across the endless sweep of the prairies.

As it grew lighter the cattle became restless, rising and stretching themselves while we continued to ride around them.

*"Then the bronc' began to pitch*
*And I began to ride;*
*He bucked me off a cut bank,*
*Hell! I nearly died!"*

sang Latigo from the other side of the herd. A yell from the wagons told that the cook was summoning the sleeping cow-punchers to breakfast; we were soon able to distinguish their figures as they rolled out of their bedding, wrapped and corded it into bundles, and huddled sullenly round the little fires. The horse-wranglers were driving in the saddle bands. All the cattle got on their feet and started feeding. In a few minutes the hasty breakfast at the wagons had evidently been despatched, for we could see the men forming rope corrals into which the ponies were driven; then each man saddled, bridled, and mounted his horse, two or three of the half-broken beasts bucking, rearing, and plunging frantically in the vain effort to unseat their riders.

The two men who were first in the saddle relieved Latigo and myself, and we immediately galloped to camp, shifted our saddles to fresh animals, gulped down a cup or two of hot coffee, and some pork, beans, and bread, and rode to the spot where the others were gathered, lolling loosely in their saddles and waiting for the round-up boss to assign them their tasks. We were the last and as soon as we arrived the boss divided all into two parties for the morning work, or "circle riding," whereby the cattle were to be gathered for the round-up proper. Then, as the others started, he turned to me and remarked: "We've got enough hands to drive this open country without you; but we're out of meat, and I don't want to kill a beef for such a small outfit; can't you shoot some antelope this morning? We'll pitch camp by the big blasted cottonwood at the foot of the ash coulées, over yonder, below the breaks of Dry Creek."

Of course I gladly assented, and was speedily riding alone across the grassy slopes. There was no lack of the game I was after, for from every rise of ground I could see antelope scattered across the prairie, singly, in couples, or in bands. But their very numbers, joined to the lack of cover on such an open, flattish country, proved a bar to success; while I was stalking one band another was sure to see me and begin running, whereat the first would likewise start; I missed one or two very long shots, and noon found me still without game.

However, I was then lucky enough to see a band of a dozen feeding to windward of a small butte, and by galloping in a long circle I got within a quarter of a mile of them before having to dismount. The stalk itself was almost too easy; for I simply walked to the butte, climbed carefully up a slope where the soil was firm and peered over the top to see the herd, a little one, a hundred yards off. They saw me at once and ran, but I held well ahead of a fine young prongbuck, and rolled him over like a rabbit, with both shoulders broken. In a few minutes I was riding onward once more, with the buck lashed behind my saddle.

The next one I got, a couple of hours later, offered a much more puzzling stalk. He was a big fellow in company with four does or small bucks. All five were lying in the middle of a slight basin, at the head of a gentle valley. At first sight it seemed impossible to get near them, for there was not so much cover as a sage-brush, and the smooth, shallow basin in which they lay was over a thousand yards across, while they were looking directly down the valley. However, it is curious how hard it is to tell, even from near by, whether a stalk can or cannot be made; the difficulty being to estimate the exact amount of shelter yielded by little inequalities of ground. In this instance a small, shallow watercourse, entirely dry, ran along the valley, and after much study I decided to try to crawl up it, although the big bulging telescopic eyes of the prongbuck—which have much keener sight than deer or any other game—would in such case be pointed directly my way.

Having made up my mind I backed cautiously down from the coign of vantage whence I had first seen the game, and ran about a mile to the mouth of a washout which formed the continuation of the watercourse in question. Protected by the high clay banks of this washout I was able to walk upright until within half a mile of the prongbucks; then my progress became very tedious and toilsome, as I had to work my way up the watercourse flat on my stomach, dragging the rifle beside me. At last I reached a spot beyond which not even a snake could crawl unnoticed. In front was a low bank a couple of feet high, crested with tufts of coarse grass. Raising my head very cautiously I peered through these and saw the prong-horn about a hundred and fifty yards distant. At the same time I found that I had crawled to the edge of a village of prairie-dogs, which had already made me aware of their presence by their shrill yelping. They saw me at once; and all those away from their homes scuttled toward them, and dived down the burrows, or sat on the mounds at the entrances, scolding convulsively and jerking their fat little bodies and short tails. This commotion at once attracted the attention of the antelope. They rose forthwith, and immediately caught a glimpse of the black muzzle of the rifle, which I was gently pushing through the grass tufts. The fatal curiosity which so often in this species offsets wariness and sharp sight proved my friend; evidently the antelope could not quite make me out and wished to know what I was. They moved nervously to and fro, striking the earth with their fore hoofs, and now and then uttering a sudden bleat. At last the big buck stood still broadside to me and I fired. He went off with the others, but lagged behind as they passed over the hill crest, and when I reached it I saw him standing, not very far off, with his head down. Then he walked backward a few steps, fell over on his side, and died.

As he was a big buck I slung him across the saddle, and started for camp afoot, leading the horse. However, my hunt was not over, for while still a mile from the wagons, going down a coulée of Dry Creek, a yearling prongbuck walked over the divide to my right and stood still until I sent a bullet into its chest; so that I made my appearance in camp with three antelope.

I spoke above of the sweet singing of the Western meadow-lark and plains skylark; neither of them kin to the true skylark, by the way, one being a cousin of the grackles and hangbirds, and the other a kind of pipit. To me both of these birds are among the most attractive singers to which I have ever listened; but with all bird music much must be allowed for the surroundings and much for the mood and the keenness of sense of the listener. The lilt of the little plains skylark is neither very powerful nor very melodious; but it is sweet, pure, long-sustained, with a ring of courage befitting a song uttered in highest air.

The meadow-lark is a singer of a higher order, deserving to rank with the best. Its song has length, variety, power, and rich melody; and there is in it sometimes a cadence of wild sadness, inexpressibly touching. Yet I cannot say that either song would appeal to others as it appeals to me; for to me it comes forever laden with a hundred memories and associations; with the sight of dim hills reddening in the dawn, with the breath of cool morning winds blowing across lonely plains, with the scent of flowers on the sunlit prairie, with the motion of fiery horses, with all the strong thrill of eager and buoyant life. I doubt if any man can judge dispassionately the bird songs of his own country; he cannot disassociate them from the sights and sounds of the land that is so dear to him.

This is not a feeling to regret, but it must be taken into account in accepting any estimate of bird music—even in considering the reputation of the European skylark and nightingale. To both of these birds I have often listened in their own homes; always with pleasure and admiration, but always with a growing belief that relatively to some other birds they were ranked too high. They are pre-eminently birds with literary associations; most people take their opinions of them at second hand, from the poets.

No one can help liking the lark; it is such a brave, honest, cheery bird, and, moreover, its song is uttered in the air, and is very long-sustained. But it is by no means a musician of the first rank. The nightingale is a performer of a very different and far higher order; yet, though it is indeed a notable and admirable singer, it is an exaggeration to call it unequalled. In melody, and above all in that finer, higher melody where the chords vibrate with the touch of eternal sorrow, it cannot rank with such singers as the wood-thrush and hermit-thrush. The serene, ethereal beauty of the hermit's song, rising and falling through the still evening under the archways of hoary mountain forests that have endured from time everlasting; the golden, leisurely chiming of the wood-thrush, sounding on June afternoons, stanza by stanza through sun-flecked groves of tall hickories, oaks, and chestnuts—with these there is nothing in the nightingale's song to compare. But in volume and continuity; in tuneful, voluble, rapid outpouring and ardor; above all, in skillful and intricate variation of theme, its song far surpasses that of either of the thrushes. In all these respects it is more just to compare it with the mocking-bird's, which, as a rule, likewise falls short precisely on those points where the songs of the two thrushes excel.

The mocking-bird is a singer that has suffered much in reputation from its powers of mimicry. On ordinary occasions, and especially in the daytime, it insists on playing the harlequin. But when free in its own favorite haunts at night in the love season it has a song, or rather songs, which are not only purely original but are also more beautiful than any other bird music whatsoever. Once I listened to a mocking-bird singing the livelong spring night, under the full moon, in a magnolia tree; and I do not think I shall ever forget its song.

It was on the plantation of Major Campbell Brown, near Nashville, in the beautiful, fertile mid-Tennessee country. The mocking-birds were prime favorites on the place; and were given full scope for the development not only of their bold friendliness toward mankind but also of that marked individuality and originality of character in which they so far surpass every other bird as to become the most interesting of all feathered folk. One of the mockers, which lived in the hedge bordering the garden, was constantly engaged in an amusing feud with an honest old setter dog, the point of attack being the tip of the dog's tail. For some reason the bird seemed to regard any hoisting of the setter's tail as a challenge and insult. It would flutter near the dog as he walked; the old setter would become interested in something and raise his tail. The bird would promptly fly at it and peck the tip; whereupon down went the tail until in a couple of minutes the old fellow would forget himself, and the scene would be repeated. The dog usually bore the assaults with comic resignation; and the mocker easily avoided any momentary outburst of clumsy resentment.

On the evening in question the moon was full. My host kindly assigned me a room of which the windows opened on a great magnolia-tree, where, I was told, a mocking-bird sang every night and all night long. I went to my room about ten. The moonlight was shining in through the open window, and the mocking-bird was already in the magnolia. The great tree was bathed in a flood of shining silver; I could see each twig and mark every action of the singer, who was pouring forth such a rapture of ringing melody as I have never listened to before or since. Sometimes he would perch motionless for many minutes, his body quivering and thrilling with the outpour of music. Then he would drop softly from twig to twig, until the lowest limb was reached, when he would rise, fluttering and leaping through the branches, his song never ceasing for an instant, until he reached the summit of the tree and launched into the warm, scent-laden air, floating in spirals, with outspread wings, until, as if spent, he sank gently back into the tree and down through the branches, while his song rose into an ecstasy of ardor and passion. His voice rang like a clarionet, in rich, full tones, and his execution covered the widest possible compass; theme followed theme, a torrent of music, a swelling tide of harmony, in which scarcely any two bars were alike. I stayed till midnight listening to him; he was singing when I went to sleep; he was still singing when I woke a couple of hours later; he sang through the livelong night.

There are many singers beside the meadow-lark and little skylark in the plains country; that brown and desolate land, once the home of the thronging buffalo, still haunted by the bands of the prongbuck, and roamed over in ever-increasing numbers by the branded herds of the ranchman. In the brush of the river bottoms there are the thrasher and song-sparrow; on the grassy uplands the lark-finch, vesper-sparrow, and lark-bunting; and in the rough canyons the rock-wren, with its ringing melody.

Yet in certain moods a man cares less for even the loveliest bird songs than for the wilder, harsher, stronger sounds of the prairie-fowl and the great sage-fowl in spring; the honking of gangs of wild geese, as they fly in rapid wedges; the bark of an eagle, wheeling in the shadow of storm-scarred cliffs; or the far-off clanging of many sand-hill cranes, soaring high overhead in circles which cross and recross at an incredible altitude. Wilder yet, and stranger, are the cries of the great four-footed beasts; the rhythmic pealing of a bull elk's challenge; and that most sinister and mournful sound, ever fraught with foreboding of murder and rapine, the long-drawn baying of the gray wolf.

Indeed, save to the trained ear, most mere bird songs are not very noticeable. The ordinary wilderness-dweller, whether hunter or cowboy, scarcely heeds them; and in fact knows but little of the smaller birds. If a bird has some conspicuous peculiarity of look or habit he will notice its existence; but not otherwise. He knows a good deal about magpies, whiskey-jacks, or water-ousels; but nothing whatever concerning the thrushes, finches, and warblers.

It is the same with mammals. The prairie-dogs he cannot help noticing. With the big pack-rats also he is well acquainted; for they are handsome, with soft gray fur, large eyes, and bushy tails; and, moreover, no one can avoid remarking their extraordinary habit of carrying to their burrows everything bright, useless, and portable, from an empty cartridge-case to a skinning-knife. But he knows nothing of mice, shrews, pocket-gophers, or weasels; and but little even of some larger mammals with very marked characteristics. Thus I have met but one or two plainsmen who knew anything of the curious plains ferret, that rather rare weasel-like animal which plays the same part on the plains that the mink does by the edges of all our streams and brooks and the tree-loving sable in the cold northern forests. The ferret makes its home in burrows, and by preference goes abroad at dawn and dusk, but sometimes even at midday. It is as bloodthirsty as the mink itself, and its life is one long ramble for prey, gophers, prairie-dogs, sage-rabbits, jack-rabbits, snakes, and every kind of ground-bird furnishing its food. I have known one to fairly depopulate a prairie-dog town, it being the arch-foe of these little rodents, because of its insatiable blood lust and its capacity to follow them into their burrows. Once I found the bloody body and broken eggs of a poor prairie-hen which a ferret had evidently surprised on her nest. Another time one of my men was eye-witness to a more remarkable instance of the little animal's bloodthirsty ferocity. He was riding the range, and being attracted by a slight commotion in a clump of grass, he turned his horse thither to look, and to his astonishment found an antelope fawn at the last gasp, but still feebly struggling, in the grasp of a ferret, which had throttled it and was sucking its blood with hideous greediness. He avenged the murdered innocent by a dexterous blow with the knotted end of his lariat.

That mighty bird of rapine, the war-eagle, which on the great plains and among the Rockies supplants the bald-headed eagle of better-watered regions, is another dangerous foe of the young antelope. It is even said that under exceptional circumstances eagles will assail a full-grown pronghorn; and a neighboring ranchman informs me that he was once an eye-witness to such an attack. It was a bleak day in the late winter, and he was riding home across a wide, dreary plateau, when he saw two eagles worrying and pouncing on a prongbuck—seemingly a yearling. It made a gallant fight. The eagles hovered over it with spread wings, now and then swooping down, their talons outthrust, to strike at the head, or to try to settle on the loins. The antelope reared and struck with hoofs and horns like a goat; but its strength was failing rapidly, and doubtless it would have succumbed in the end had not the approach of the ranchman driven off the marauders.

I have likewise heard stories of eagles attacking badgers, foxes, bobcats, and coyotes; but I am inclined to think all such cases exceptional. I have never myself seen an eagle assail anything bigger than a fawn, lamb, kid, or jack-rabbit. It also swoops at geese, sage-fowl, and prairie-fowl. On one occasion while riding over the range I witnessed an attack on a jack-rabbit. The eagle was soaring overhead, and espied the jack while the latter was crouched motionless. Instantly the great bird rushed down through the humming air, with closed wings; checked itself when some forty yards above the jack, hovered for a moment, and again fell like a bolt. Away went long-ears, running as only a frightened jack can; and after him the eagle, not with the arrowy rush of its descent from high air, but with eager, hurried flapping. In a short time it had nearly overtaken the fugitive when the latter dodged sharply to one side, and the eagle overshot it precisely as a greyhound would have done, stopping itself by a powerful, setting motion of the great pinions. Twice this manœver was repeated; then the eagle made a quick rush, caught and overthrew the quarry before it could turn, and in another moment was sitting triumphant on the quivering body, the crooked talons driven deep into the soft, furry sides.

Once while hunting mountain-sheep in the Bad Lands I killed an eagle on the wing with the rifle. I was walking beneath a cliff of gray clay, when the eagle sailed into view over the crest. As soon as he saw me he threw his wings aback, and for a moment before wheeling poised motionless, offering a nearly stationary target; so that my bullet grazed his shoulder, and down he came through the air, tumbling over and over. As he struck the ground he threw himself on his back, and fought against his death with the undaunted courage proper to his brave and cruel nature.

Indians greatly prize the feathers of this eagle. With them they make their striking and beautiful war-bonnets and bedeck the manes and tails of their spirited war ponies. Every year the Grosventres and Mandans from the Big Missouri come to the neighborhood of my ranch to hunt. Though not good marksmen, they kill many whitetail deer, driving the bottoms for them in bands, on horseback; and they catch many eagles. Sometimes they take these alive by exposing a bait near which a hole is dug, where one of them lies hidden for days, with Indian patience, until an eagle lights on the bait and is noosed.

Even eagles are far less dangerous enemies to antelope than are wolves and coyotes. These beasts are always prowling round the bands to snap up the sick or unwary; and in spring they revel in carnage of the kids and fawns. They are not swift enough to overtake the grown animals by sheer speed; but they are superior in endurance and, especially in winter, often run them down in fair chase. A prongbuck is a plucky little beast and when cornered it often makes a gallant, though not a very effectual, fight.

# 14

# My Life as a Naturalist

I AM ASKED TO GIVE AN ACCOUNT OF MY INTEREST IN NATURAL history, and my experience as an amateur naturalist. The former has always been very real; and the latter, unfortunately, very limited.

I don't suppose that most men can tell why their minds are attracted to certain studies any more than why their tastes are attracted by certain fruits. Certainly, I can no more explain why I like "natural history" than why I like California canned peaches; nor why I do not care for that enormous brand of natural history which deals with invertebrates any more than why I do not care for brandied peaches. All I can say is that almost as soon as I began to read at all I began to like to read about the natural history of the beasts and birds and the more formidable or interesting reptiles and fishes.

The fact that I speak of "natural history" instead of "biology," and use the former expression in a restricted sense, will show that I am a belated member of the generation that regarded Audubon with veneration, that accepted Waterton—Audubon's violent critic—as the ideal of the wandering naturalist, and that looked upon Brehm as a delightful but rather awesomely erudite example of advanced scientific thought. In the broader field, thank Heaven, I sat at the feet of Darwin and Huxley; and studied the large volumes in which Marsh's and Leidy's paleontological studies were embalmed with a devotion that was usually attended by a dreary lack of reward—what would I not have given fifty years ago for a writer like Henry Fairfield Osborn, for some scientist who realized that intelligent laymen need a guide capable of building before their eyes the life that was, instead of merely cataloguing the fragments of the death that is.

I was a very near-sighted small boy, and did not even know that my eyes were not normal until I was fourteen; and so my field studies up to that period were even more worthless than those of the average boy who "collects" natural-history specimens much as he collects stamps. I studied books industriously but nature only so far as could be compassed by a mole-like vision; my triumphs consisted in such things as bringing home and raising—by the aid of milk and a syringe—a family of very young gray squirrels, in fruitlessly endeavoring to tame an excessively unamiable woodchuck, and in making friends with a gentle, pretty, trustful white-footed mouse which reared her family in an empty flower-pot. In order to attract my attention birds had to be as conspicuous as bobolinks or else had to perform feats such as I remember the barn-swallows of my neighborhood once performed, when they assembled for the migration alongside our house and because of some freak of bewilderment swarmed in through the windows and clung helplessly to the curtains, the furniture, and even to our clothes.

Just before my fourteenth birthday my father—then a trustee of the American Museum of Natural History—started me on my rather moth-like career as a naturalist by giving me a pair of spectacles, a French pin-fire double-barrelled shotgun—and lessons in stuffing birds. The spectacles literally opened a new world to me. The mechanism of the pin-fire gun was without springs and therefore could not get out of order—an important point, as my mechanical ability was nil. The lessons in stuffing and mounting birds were given me by Mr. John G. Bell, a professional taxidermist and collector who had accompanied Audubon on his trip to the then "far West." Mr. Bell was a very interesting man, an American of the before-the-war type. He was tall, straight as an Indian, with white hair and smooth-shaven clear-cut face; a dignified figure, always in a black frock coat. He had no scientific knowledge of birds or mammals; his interest lay merely in collecting and preparing them. He taught me as much as my limitations would allow of the art of preparing specimens for scientific use and of mounting them. Some examples of my wooden methods of mounting birds are now in the American Museum: three different species of Egyptian plover, a snowy owl, and a couple of spruce-grouse mounted on a shield with a passenger pigeon—the three latter killed in Maine during my college vacations.

With my spectacles, my pin-fire gun, and my clumsy industry in skinning "specimens," I passed the winter of '72–'73 in Egypt and Palestine, being then fourteen years old. My collections showed nothing but enthusiasm on my part. I got no bird of any unusual scientific value. My observations were as valueless as my collections save on just one small point; and this point is of interest only as showing, not my own power of observation, but the ability of good men to fail to observe or record the seemingly self-evident.

On the Nile the only book dealing with Egyptian birds which I had with me was one by an English clergyman, a Mr. Smith, who at the end of his second volume gave a short list of the species he has shot, with some comments on their habits, but without descriptions. On my way home through Europe I secured a good book of Egyptian ornithology by a Captain Shelley. Both books enumerated and commented on several species of chats—the Old World chats, of course, which have nothing in common with our queer warbler of the same name. Two of these chats were common along the edges of the desert. One species was a boldly pied black-and-white bird, the other was colored above much like the desert sand, so that when it crouched it was hard to see. I found that the strikingly conspicuous chat never tried to hide, was very much on the alert, and was sure to attract attention when a long way off; whereas the chat whose upper color harmonized with its surroundings usually sought to escape observation by crouching motionless. These facts were obvious even to a dull-sighted, not particularly observant boy; they were essential features in the comparison between and in the study of the life histories of the two birds. Yet neither of the two books in my possession so much as hinted at them.

I think it was my observation of these, and a few similar facts, which prevented my yielding to the craze that fifteen or twenty years ago became an obsession with certain otherwise good men—

the belief that all animals were protectively colored when in their natural surroundings. That this simply wasn't true was shown by a moment's thought of these two chats; no rational man could doubt that one was revealingly and the other concealingly colored; and each was an example of what was true in thousands of other cases. Moreover, the incident showed the only, and very mild, merit which I ever developed as a "faunal naturalist." I never grew to have keen powers of observation. But whatever I did see I saw truly, and I was fairly apt to understand what it meant. In other words, I saw what was sufficiently obvious, and in such case did not usually misinterpret what I had seen. Certainly this does not entitle me to any particular credit, but the outstanding thing is that it does entitle me to some, even although of a negative kind; for the great majority of observers seem quite unable to see, to record, or to understand facts so obvious that they leap to the eye. My two ornithologists offered a case in point as regards the chats; and I shall shortly speak of one or two other cases, as, for example, the cougar and the saddle-backed lechwe.

After returning to this country and until I was half-way through college, I continued to observe and collect in the fashion of the ordinary boy who is interested in natural history. I made copious and valueless notes. As I said above, I did not see and observe very keenly; later it interested and rather chagrined me to find out how much more C. Hart Merriam and John Burroughs saw when I went out with them near Washington or in the Yellowstone Park; or how much more George K. Cherrie and Leo E. Miller and Edmund Heller and Edgar A. Mearns and my own son Kermit saw in Africa and South America, on the trips I took to the Nyanza lakes and across the Brazilian hinterland.

During the years when I was a boy I "collected specimens" at Oyster Bay or in the North woods, my contributions to original research were of minimum worth—they were limited to occasional records of such birds as the dominica warbler at Oyster Bay, or to seeing a duck-hawk work havoc in a loose gang of night-herons, or to noting the bloodthirsty conduct of a captive mole-shrew—I think I sent an account of the last incident to C. Hart Merriam. I occasionally sent to some small ornithological publication a local list of Adirondack birds or something of the sort; and then proudly kept reprinted copies of the list on my desk until they grew dog-eared and then disappeared. I lived in a region zoologically so well known that the obvious facts had all been set forth already, and as I lacked the power to find out the things that were not obvious, my work merely paralleled the similar work of hundreds of other young collectors who had a very good time but who made no particular addition to the sum of human knowledge.

Among my boy friends who cared for ornithology was a fine manly young fellow, Fred Osborn, the brother of Henry Fairfield Osborn. He was drowned, in his gallant youth, forty years ago; but he comes as vividly before my eyes now as if he were still alive. One cold and snowy winter I spent a day with him at his father's house at Garrison-on-the-Hudson. Numerous northern birds, which in our eyes were notable rarities, had come down with the hard weather. I spied a flock of crossbills in a pine, fired, and excitedly rushed forward. A twig caught my spectacles and snapped them I knew not where. But dim though my vision was, I could still make out the red birds lying on the snow; and to me they were treasures of such importance that I abandoned all thought of my glasses and began a near-sighted hunt for my quarry. By the time I had picked up the last crossbill I found that I had lost all trace of my glasses; my day's sport—or scientific endeavor, whatever you choose to call it—came to an abrupt end; and as a result of the lesson I never again in my life went out shooting, whether after sparrows or elephants, without a spare pair of spectacles in my pocket. After some ranch experiences I had my spectacle cases made of steel; and it was one of these spectacle cases which saved my life in after-years when a man shot into me in Milwaukee.

While in Harvard I was among those who joined in forming the Nuttall Club, which I believe afterward became one of the parent sources of the American Ornithologists' Union.

The Harvard of that day was passing through a phase of biological study which was shaped by the belief that German university methods were the only ones worthy of copy, and also by the proper admiration for the younger Agassiz, whose interest was mainly in the lower forms of marine life. Accordingly it was the accepted doctrine that a biologist—the word "naturalist" was eschewed as archaic—was to work toward the ideal of becoming a section-cutter of tissue, who spent his time studying this tissue, and low marine organisms, under the microscope. Such work was excellent; but it covered a very small part of the biological field; and not only was there no encouragement for the work of the field naturalist, the faunal naturalist, but this work was positively discouraged, and was treated as of negligible value. The effect of this attitude, common at that time to all our colleges, was detrimental to one very important side of natural-history research. The admirable work of the microscopist had no attraction for me, nor was I fitted for it; I grew even more interested in other forms of work than in the work of a faunal naturalist; and I abandoned all thought of making the study of my science my life interest.

But I never lost a real interest in natural history; and I very keenly regret that at certain times I did not display this interest in more practical fashion. Thus, for the dozen years beginning with 1883, I spent much of my time on the Little Missouri, where big game was then plentiful. Most big-game hunters never learn anything about the game except how to kill it; and most naturalists never observe it at all. Therefore a large amount of important and rather obvious facts remain unobserved or inaccurately observed until the species becomes extinct. What is most needed is not the ability to see what very few people can see, but to see what almost anybody can see, but nobody takes the trouble to look at. But I vaguely supposed that the obvious facts were known; and I let most of the opportunities pass by. Even so, many of my observations on the life histories of the bighorns, white goats, prongbucks, deer, and wapiti, and occasional observations on some of the other beasts, such as black-footed ferrets, were of value; indeed as regards some of the big-game beasts, the account in "Hunting Trips of a Ranchman," "Ranch Life and the Hunting Trail," and "The Wilderness Hunter" gave a good deal of information which, as far as I know, is not to be found elsewhere.

To illustrate what I mean as "obvious" facts which nevertheless are of real value I shall instance the cougar. In the winter of 1901 I made a cougar-hunt with hounds, spending about five weeks in the mountains of northwestern Colorado. At that time the cougar had been seemingly well known to hunters, settlers, naturalists, and novelists for more than a century; and yet it was actually impossible to get trustworthy testimony on such elementary points as, for instance, whether the male and female mated permanently, or at least until the young were reared (like foxes and wolves), and whether the animal caught its prey by rambling and stalking or, as was frequently asserted, by lying in wait on the branches of a tree. The facts I saw and observed during our five weeks' hunt in the snow were obvious; they needed only the simplest powers of observation and of deduction from observation. But nobody had hitherto shown or exercised these simple powers! My narrative in the volume "Outdoor Pastimes of an American Hunter" gave the first reasonably full and trustworthy life history of the cougar as regards its most essential details—for Merriam's capital Adirondack study had dealt with the species when it was too near the vanishing-point and therefore when the conditions were too abnormal for some of these essential details to be observed.

In South America I made observations of a certain value on some of the strange creatures we met, and these are to be found in the volume "Through the Brazilian Wilderness"; but the trip was primarily one of exploration. In Africa, however, we really did some good work in natural history. Many of my observations were set forth in my book "African Game Trails"; and I have always felt that the book which Edmund Heller and I jointly wrote, the "Life Histories of African Game Animals," was a serious and worth-while contribution to science. Here again, this contribution, so far as I was concerned, consisted chiefly in seeing, recording, and interpreting facts which were really obvious, but to which observers hitherto had been blind, or which they had misinterpreted partly because sportsmen seemed incapable of seeing anything except as a trophy, partly because stay-at-home systematists never saw anything at all except skins and skulls which enabled them to give Latin names to new "species" or "sub-species," partly because collectors had collected birds and beasts in precisely the spirit in which other collectors assembled postage-stamps.

I shall give a few instances. In mid-Africa we came across a peculiar bat, with a greenish body and slate-blue wings. Specimens of this bat had often been collected. But I could find no record of its really interesting habits. It was not nocturnal; it was hardly even crepuscular. It hung from the twigs of trees during the day and its activities began rather early in the afternoon. It did not fly continuously in swallow fashion, according to the usual bat custom. It behaved like a phœbe or other flycatcher. It hung from a twig until it saw an insect, then swooped down, caught the insect, and at once returned to the same or another twig—just as a phœbe or pewee or king-bird returns to its perch after a similar flight.

On the White Nile I hunted a kind of handsome river antelope, the white-withered or saddle-backed lechwe. It had been known for fifty years to trophy-seeking sportsmen, and to closet naturalists, some of whom had called it a kob and others a water-buck. Its nearest kinsman was in reality the ordinary lechwe, which dwelt far off to the south, along the Zambezi. But during that half-century no hunter or closet naturalist had grasped this obvious fact. I had never seen the Zambezi lechwe, but I had carefully read the account of its habits by Selous—a real hunter-naturalist, faunal naturalist. As soon as I came across the White Nile river bucks, and observed their habits, I said to my companions that they were undoubtedly lechwes; I wrote this to Selous, and to another English hunter-naturalist, Migand; and even a slight examination of the heads and skins when compared with those of the other lechwe and of the kobs and water-bucks proved that I was right.

A larger, but equally obvious group of facts was that connected with concealing and revealing coloration. As eminent a naturalist as Wallace, and innumerable men of less note, had indulged in every conceivable vagary of speculative theory on the subject, largely based on supposed correlation between the habits and the shape or color patterns of big animals which, as a matter of fact, they had never seen in a state of nature. While in Africa I studied the question in the field, observing countless individuals of big beasts and birds, and comparing the results with what I had observed of the big game and the birds of North America (the result being borne out by what I later observed in South America). In a special chapter of the "Life Histories of African Game Animals," as well as in a special number of the *American Museum Bulletin*, I set forth the facts thus observed and the conclusions inevitably to be deduced from them. All that I thus set forth, and all the conclusions I deduced, belonged to the obvious; but that there was need of thus setting forth the obvious was sufficiently shown by the simple fact that large numbers of persons refused to accept it even when set forth.

I do not think there is much else for me to say about my anything but important work as a naturalist. But perhaps I may say further that while my interest in natural history has added very little to my sum of achievement, it has added immeasurably to my sum of enjoyment in life.

# Notes

### 3. THE BIG-HORN SHEEP

1. I have camped out when the thermometer showed 65 degrees of frost; *not* –*65°*, as I see I once put it by mistake in copying my rough field-notes.

2. Actual pacing, not guesswork.

### 5. DOWN AN UNKNOWN RIVER

1. The first four days, before we struck the upper rapids, and during which we made nearly seventy kilometres, are of course not included when I speak of our making our way down the rapids.

2. The above account of all the circumstances connected with the murder was read to and approved as correct by all members of the expedition.

3. I hope that this year Ananás, or Pineapple, will also be put on the map. One of Colonel Rondon's subordinates is to attempt the descent of the river. We passed the headwaters of the Pineapple on the high plateau, very possibly we passed its mouth, although it is also possible that it empties into the Canumá or Tapajos. But it will not be "put on the map" until some one descends and finds out where, as a matter of fact, it really does go.

### 9. BIRD RESERVES

1. An expression borrowed from Stewart Edward White's capital "Rediscovered Country."

### 12. THE WAPITI OR ROUND-HORNED ELK

1. Steps in the direction indicated are now being taken by the Federal authorities.

# Sources of the Readings

FOR THOSE SEEKING DEFINITIVE BIBLIOGRAPHICAL DETAILS on TR's books I can't recommend too highly the long-needed and invaluable reference work by Heather G. Cole and R. W. G. Vail, *Theodore Roosevelt: A Descriptive Biography* (New Castle DE: Oak Knoll Press, 2020).

## PART ONE, WILDERNESS ADVENTURES

### 1. THE JOY OF THE WILD

"The Joy of the Wild." Foreword to *A Book-Lover's Holidays in the Open*, vii–x. New York: Charles Scribner's Sons, 1916.

### 2. THE AMERICAN WILDERNESS

"The American Wilderness: Wilderness Hunters and Wilderness Game." In *The Wilderness Hunter*, 1–19. New York: G. P. Putnam's Sons, 1893.

### 3. THE BIG-HORN SHEEP

"The Big-Horn Sheep." In *Ranch Life and the Hunting Trail*, 153–70. New York: Century, 1888.

### 4. IN THE LOUISIANA CANEBRAKES

"In the Louisiana Canebrakes." *Scribner's Magazine*, January 1908.

### 5. DOWN AN UNKNOWN RIVER

"Down an Unknown River into the Equatorial Forest." In *Through the Brazilian Wilderness*, 282–320. New York: Charles Scribner's Sons, 1914.

### PART TWO, WILDERNESS PRESERVATION

#### 6. A NATIONAL PARK SERVICE

"A National Park Service." *The Outlook*, February 3, 1912.

#### 7. JOHN MUIR

"John Muir: An Appreciation." *The Outlook*, January 6, 1915.

#### 8. WILDERNESS RESERVES

"Wilderness Reserves: The Yellowstone Park." In *Outdoor Pastimes of an American Hunter*, 287–318. New York: Charles Scribner's Sons, 1905.

#### 9. BIRD RESERVES

"Bird Reserves at the Mouth of the Mississippi." In *A Book-Lover's Holidays in the Open*, 274–317. New York: Charles Scribner's Sons, 1916.

#### 10. GRAND CANYON SPEECH

"Speech at the Grand Canyon, May 6, 1903." *New York Sun*, May 7, 1903.

### PART THREE, NATURAL HISTORY

#### 11. SMALL COUNTRY NEIGHBORS

"Small Country Neighbors." *Scribner's Magazine*, October 1907.

#### 12. THE WAPITI OR ROUND-HORNED ELK

"The Wapiti or Round-Horned Elk." In *The Deer Family*, 131–64. Edited by Theodore Roosevelt, Theodore S. Van Dyke, Daniel Giraud Elliot, and A. J. Stone. New York: MacMillan, 1902. Note that TR used this same title for an earlier essay with different content in *The Wilderness Hunter*, chapter 9, 156–76.

#### 13. ON THE CATTLE RANGES

"On the Cattle Ranges." In *The Wilderness Hunter*, 55–73. New York: G. P. Putnam's Sons, 1893.

#### 14. MY LIFE AS A NATURALIST

"My Life as a Naturalist." *American Museum Journal*, May 1918.

www.ingramcontent.com/pod-product-compliance
Lightning Source LLC
Chambersburg PA
CBHW021256090325
23029CB00014B/86
*9781496240521*